Minerals: A Very Short Introduction

VERY SHORT INTRODUCTIONS are for anyone wanting a stimulating and accessible way in to a new subject. They are written by experts, and have been translated into more than 40 different languages.

The Series began in 1995, and now covers a wide variety of topics in every discipline. The VSI library now contains over 350 volumes—a Very Short Introduction to everything from Psychology and Philosophy of Science to American History and Relativity—and continues to grow in every subject area.

Very Short Introductions available now:

ACCOUNTING Christopher Nobes
ADVERTISING Winston Fletcher
AFRICAN AMERICAN RELIGION
    Eddie S. Glaude Jr.
AFRICAN HISTORY John Parker and
    Richard Rathbone
AFRICAN RELIGIONS
    Jacob K. Olupona
AGNOSTICISM Robin Le Poidevin
ALEXANDER THE GREAT
    Hugh Bowden
AMERICAN HISTORY Paul S. Boyer
AMERICAN IMMIGRATION
    David A. Gerber
AMERICAN LEGAL HISTORY
    G. Edward White
AMERICAN POLITICAL PARTIES
    AND ELECTIONS L. Sandy Maisel
AMERICAN POLITICS
    Richard M. Valelly
THE AMERICAN PRESIDENCY
    Charles O. Jones
AMERICAN SLAVERY
    Heather Andrea Williams
ANAESTHESIA Aidan O'Donnell
ANARCHISM Colin Ward
ANCIENT EGYPT Ian Shaw
ANCIENT EGYPTIAN ART AND
    ARCHITECTURE Christina Riggs
ANCIENT GREECE Paul Cartledge
THE ANCIENT NEAR EAST
    Amanda H. Podany
ANCIENT PHILOSOPHY Julia Annas
ANCIENT WARFARE
    Harry Sidebottom

ANGELS David Albert Jones
ANGLICANISM Mark Chapman
THE ANGLO-SAXON AGE John Blair
THE ANIMAL KINGDOM
    Peter Holland
ANIMAL RIGHTS David DeGrazia
THE ANTARCTIC Klaus Dodds
ANTISEMITISM Steven Beller
ANXIETY Daniel Freeman and
    Jason Freeman
THE APOCRYPHAL GOSPELS
    Paul Foster
ARCHAEOLOGY Paul Bahn
ARCHITECTURE Andrew Ballantyne
ARISTOCRACY William Doyle
ARISTOTLE Jonathan Barnes
ART HISTORY Dana Arnold
ART THEORY Cynthia Freeland
ASTROBIOLOGY David C. Catling
ATHEISM Julian Baggini
AUGUSTINE Henry Chadwick
AUSTRALIA Kenneth Morgan
AUTISM Uta Frith
THE AVANT GARDE David Cottington
THE AZTECS David Carrasco
BACTERIA Sebastian G. B. Amyes
BARTHES Jonathan Culler
THE BEATS David Sterritt
BEAUTY Roger Scruton
BESTSELLERS John Sutherland
THE BIBLE John Riches
BIBLICAL ARCHAEOLOGY
    Eric H. Cline
BIOGRAPHY Hermione Lee
THE BLUES Elijah Wald

Available soon:

For more information visit our website

www.oup.com/vsi/

David J. Vaughan

MINERALS

A Very Short Introduction

OXFORD
UNIVERSITY PRESS

OXFORD
UNIVERSITY PRESS

Great Clarendon Street, Oxford, OX2 6DP,
United Kingdom

Oxford University Press is a department of the University of Oxford.
It furthers the University's objective of excellence in research, scholarship,
and education by publishing worldwide. Oxford is a registered trade mark of
Oxford University Press in the UK and in certain other countries

© David J. Vaughan 2014

The moral rights of the author have been asserted

First edition published in 2014

Published in the United States of America by Oxford University Press
198 Madison Avenue, New York, NY 10016, United States of America

British Library Cataloguing in Publication Data
Data available

Library of Congress Control Number: 2014942528

ISBN 978-0-19-968284-3

Printed and bound by
CPI Group (UK) Ltd, Croydon, CR0 4YY

To Jane, with love

Contents

Acknowledgements

For offering critical comments on the whole volume, I wish to thank Jack Zussman, Ian Parsons, and my wife Jane. Valuable comments on particular sections were also provided by Ernie Rutter, Alison Pawley and Kate Brodie. Richard Hartley is thanked for help in drafting the figures.

List of illustrations

Minerals

The publisher and author apologize for any errors or omissions in this list. If contacted they will be happy to rectify these at the earliest opportunity.

List of illustrations

Chapter 1
The mineral world

In 2004, I had the honour of being the lecturer at a 'Friday Evening Discourse' at the Royal Institution in London. These public lectures, aimed at the popularization of science, were initiated by Michael Faraday in 1826, and were probably the first such lectures anywhere in the world. My lecture was entitled 'Minerals, Molecules and Maladies'; and, indeed, my first word was 'minerals'. I pointed out that the first experience many of us have of minerals comes from those seen in pebbles collected on a beach during a seaside holiday, or the fine specimens of minerals on display at a local museum. Although the study of minerals begins with their collection and identification, my lecture was about the new developments in what I would describe as 'modern mineralogy', the central theme of this book.

The study of minerals (mineralogy) is the most fundamental aspect of the Earth and environmental sciences. Minerals existed long before any forms of life. They have played an important role in the origin and evolution of life, and interact with biological systems in ways we are only now beginning to understand. Mineralogy is arguably also the oldest of all of the practical sciences. The first manufacture of fire that could be called upon as needed depended, in part, on the sparks produced on striking minerals such as pyrite. Although we cannot be sure of the date when the first mineral-based fire strikers were used, the earliest

1

generally accepted evidence of fire production dates back 500,000 years to *Homo erectus* populations, and *Homo sapiens* was an expert fire starter by 40,000 to 50,000 years ago, during the Palaeolithic period.

That the use of minerals was the key to human development is shown by our use of the terms 'Stone Age', 'Bronze Age', and 'Iron Age'. Stone tools were fashioned by our ancestors more than 30,000 years ago from hard, fine-grained rocks such as flint. By 9000 BC, clay minerals were being fired to make pottery, leading to other ceramic arts such as brickmaking and glassmaking by 3500 BC. By 3000 BC copper was being extracted from its ores, as were other metals (silver, lead, zinc, antimony), some of which could be used to make alloys. The smelting methods developed to extract these metals led the way for the development of the more demanding metallurgy needed to extract the most useful of all metals, iron. (The Roman writer Pliny described iron as 'the most useful and most fatal instrument in the hand of man'.) In primitive societies, an expert on minerals (a 'mineralogist') was surely required to seek out the raw materials for the production of bronze (an alloy of copper and tin) and, later, of iron. Today, minerals are essential raw materials for our technologically advanced societies. We are surrounded by structures, machines, and devices made using mineral raw materials. It can truly be said of essential resources that 'if it cannot be grown, it must be mined'.

Traditional mineralogy has been about describing, analysing (chemically and in terms of crystal structure), naming, and classifying minerals. It has also been concerned with the properties of minerals (such as optical, magnetic, or electrical properties). Modern mineralogy is more concerned with the roles different minerals play in the plate tectonic cycle on the one hand (in the Earth's interior) and the weathering–transport–deposition cycles (at the Earth's surface) on the other, and about how minerals transform under different conditions and the rates at which they change. Where minerals interact with the living world, surfaces and interfaces are central to understanding—for example, in the roles microbes play in both mineral formation and breakdown. Minerals can also be critical for human health, providing essential nutrients or releasing poisons such as arsenic (the 'mineral–human interface'). A more indirect influence of minerals on human well-being leads us to considering the topic of mineral resources—their formation, exploitation, and scarcity.

So what exactly do we mean by 'mineral'. As with many things in the natural world, the perfect definition is elusive, but a reasonable working definition is:

> a solid material, formed by natural processes, that has a regular arrangement of its constituent atoms which sets limits to its range of chemical composition and commonly gives it a characteristic crystal shape.

Here, the 'regular arrangement of…atoms' is what we term the 'crystal structure' of a mineral, and it is commonly illustrated using the kind of 'ball and spoke' model shown for galena (PbS) in Figure 1a (as further discussed below). We should note here that a 'crystal shape' may not be obvious in some cases; the crystals may be too small to see, or not well developed, if developed at all. Also, a small number of minerals fall outside this definition, such as opal which is a form of silica lacking the 'regular arrangement of its constituent atoms' (being therefore described as 'amorphous', 'non-crystalline', or at least 'poorly crystalline'). Other substances excluded by this definition are coals and bitumens, and also amber. These have no regular atomic structures or well-defined compositional limits. Also generally excluded are materials which are liquids at room temperature such as petroleum and, of course, water (but elemental mercury, a liquid above −39°C and familiar from the mercury thermometer, is classed as a mineral). We also exclude here commonplace uses of the word mineral meaning an effervescent soft drink or a substance needed by the body for good health (as in 'vitamins and minerals').

The distinction between a mineral and a rock is also important to clarify. Rocks are generally composed of a number of different minerals. For example, granites typically contain the potassium and sodium aluminium silicate minerals of the feldspar family, along with layer silicates of the mica family and silica itself, as the mineral quartz (see Table 1). The individual mineral grains in rocks such as granites will mostly range in size from a few centimetres in diameter to a millimetre or less, and will be intimately intergrown together. Although quartzite is made up, as the name implies, almost entirely of quartz, it is a rock not a mineral. It is rocks, themselves comprising minerals, which make up ocean floors or mountain ranges. Although the analogy should not be taken too far, we can liken the Earth to a human body, with rocks the limbs or organs, and minerals the different kinds of cells. Just as living cells are the fundamental components of the body, so minerals are the fundamental components of the Earth.

There are approximately 4,400 known minerals, with new mineral species being discovered every year (adding around 50 a year to the list). Proposed new minerals have to be approved by a Commission of the International Mineralogical Association. To be approved, data on the chemical composition, basic crystallography, and certain key properties have to be submitted, along with a proposed mineral name (see Box 1).

As well as the actual mineral species, some minerals occur in such different forms that they justify varietal names. For example, amethyst, jasper, agate, chalcedony, and rose quartz are all varieties of quartz. The differences between them relate to properties such as colour or crystallinity; differences that are caused by the presence of very minor impurities, or due to formation under different conditions.

Although there are thousands of minerals, very few are commonplace. In their mineralogy textbook, Darby Dyar, Mickey Gunter, and Dennis Tasa suggest a list of 'big ten minerals'. Apart

Table 1 Mineral groups

Group	Examples
Native elements	gold (Au), copper (Cu), sulphur (S), diamond and graphite (both carbon, C)
Sulphides (including arsenides)	pyrite (FeS_2), pyrrhotite ($Fe_{1-x}S$), chalcopyrite ($CuFeS_2$), galena (PbS), sphalerite (ZnS), mackinawite (FeS), arsenopyrite (FeAsS)
Oxides (including hydroxides)	hematite (Fe_2O_3), magnetite (Fe_3O_4), rutile (TiO_2), corundum (Al_2O_3), cuprite (Cu_2O), goethite (FeOOH)
Carbonates	calcite ($CaCO_3$), magnesite ($MgCO_3$), dolomite (($Ca,Mg)CO_3$)
Sulphates	gypsum ($CaSO_4 \cdot H_2O$), barite ($BaSO_4$)
Phosphates	apatite ($CaPO_4$)
Halides	halite (NaCl), fluorite (CaF_2)
Silicates:	
Framework	quartz (SiO_2), feldspar minerals— orthoclase feldspar ($KAlSi_3O_8$), plagioclase feldspar ($CaAl_2Si_2O_8 - NaAlSi_3O_8$)
Layer	mica minerals—muscovite ($K_2Al_4Si_6Al_2O_{20}(OH,F)_4$) biotite ($K_2(Mg,Fe)_6Si_6Al_2O_{20}(OH)_4$) clay minerals—kaolinite ($Al_4Si_4O_{10}(OH)_8$)
Ring	beryl—$Be_3Al_2Si_6O_{18}$
double chain	amphibole minerals—anthophyllite (($Mg,Fe)_7Si_8O_{22}(OH,F)_2$) tremolite $Ca_2(Mg,Fe)_5Si_8O_{22}(OH)_2$
single chain	pyroxene minerals—enstatite ($MgSiO_3$), diopside (($Ca,Mg)Si_2O_6$)
island	olivine (($Mg,Fe)_2SiO_4$)

Box 1 Mineral names

Some minerals have names derived from antiquity, such as *galena* which comes from a Latin word for lead ore, and *chalcopyrite* from the greek *chalkos* meaning copper. Others have names associated with a particular property (the magnetism of *magnetite*, or the rose colour of *rhodocrosite*). However, the majority of minerals are named either after the locality where they were first discovered (*aragonite* from Aragon in Spain, *montmorillonite* from Montmorillon in France, or *mackinawite* from the Mackinaw Mine, Washington, USA) or after people. In the early years of characterizing and describing minerals, this may have been a leading cultural figure such as a writer or philosopher (*goethite* named for the German poet and philosopher, Goethe). Latterly, minerals have been named after mineralogists, in recognition of their contributions to the subject (noting that it is not acceptable for the person reporting a new mineral to name it after themselves—such a suggestion would be described by the British as 'not cricket'). For obvious reasons, the same name cannot be used for more than one mineral, and this has led to some inventiveness when the obvious name has already been used. For example, the silver mineral *smithite* was named for an eminent British mineralogist; this led to scientists later wishing to honour J. V. Smith, an equally distinguished mineralogist, to name a mineral *joesmithite*. Another hazard faced by those honoured in this way is that of having a name 'discredited'. If further work shows that the original study leading to a new mineral was flawed, the name has to be removed from the list of recognized mineral species and cannot ever be used again. The way in which we name minerals may seem arcane to the lay person, and there have been proposals for inventing whole new systems. But rather like attempts to simplify the spelling of English or introduce a new international language like Esperanto, they have never been adopted. Here I must own up to having a vested interest, in that the mineral *vaughanite* was named in my honour by Canadian scientists in 1989.

from calcite (calcium carbonate) which is the dominant mineral in limestones, and quartz (silicon dioxide) which dominates in sandstones as well as being important in rocks such as granites, the other members of this list are all silicates (a group discussed in detail below). They comprise the olivines (magnesium iron silicates), pyroxenes (calcium, magnesium, iron silicates), amphiboles (calcium, magnesium, iron silicates which also contain bonded water), the mica minerals muscovite (potassium, aluminium silicates with bonded water) and biotite (potassium, magnesium, iron, aluminium silicates with bonded water), and the main minerals of the feldspar group. The feldspars are alumino-silicates with dominantly potassium (orthoclase), sodium (albite), or calcium (anorthite) as essential constituents. Alternative lists to this one could be proposed on the basis of different criteria, such as the most economically important minerals. Those above are suggested to be the 'big ten' minerals because they are by far the most abundant components of the rocks that make up the Earth's crust and, immediately below that, the upper mantle.

As we will see in Chapter 2, the characterization of a particular mineral involves determining its chemical composition and crystallographic properties. Until the middle of the 20th century, chemical analysis required taking the pure mineral, dissolving it in acid, and applying the methods used in the 'wet chemical analysis' of that solution, a procedure familiar to all who have taken chemistry courses in school or college. Remarkable results were achieved by the countless hours of patient analytical work carried out by early practitioners. Mineral analysis was revolutionized by the invention of the *electron microprobe*, an instrument in which a beam of electrons is focused using magnetic lenses so as to strike a flat polished surface of a mineral grain, stimulating the emission of X-rays. The energies and intensities of these X-rays are characteristic of the elements present, and can be used to obtain a quantitative chemical analysis of a volume of solid material as small as a few cubic micrometres (millionths of a metre; also called 'microns').

Until the early years of the 20th century, crystallography was about observing the external symmetry of natural crystals and measuring the angles between the faces of crystals. Revolutionary advances were achieved with the discovery that the interaction between a beam of X-rays and a crystalline solid could be used to determine the arrangement of the atoms in that solid. Minerals were the first crystalline solids to be studied in this way—a path which led to the greatest scientific discovery of the 20th century, the determination of the crystal structure of DNA.

The routine identification of minerals in a 'hand specimen' (which, as the name implies, is a sample of a size that fits neatly into the hand) in the field or laboratory relies upon a variety of simply observed properties. These include crystal morphology, hardness, density, presence of crystal planes along which the mineral can be 'cleaved' or 'fractured', lustre, and, in a very few cases, mineral magnetism. Colour may be diagnostic (the blues and greens of some copper minerals) but can be misleading. Quartz, for example, can be transparent and colourless, pink, purple, yellow, or black, and there is a range of non-transparent forms with darker colours.

In the field, it may be possible to make a definite, or at least a preliminary, identification by visual observation or with a simple test. If the mineral occurs as well-developed crystals, it should be possible to use the presence of symmetry elements to assign it to one of the seven crystal systems (see below). Some minerals have distinctive crystal shapes known as *habits* which may be, for example, needle-like, fibrous, platey, tabular, etc. The hardness of a mineral can be simply assessed as to whether it can scratch or be scratched by a fingernail, a copper coin, or a steel penknife blade. Minerals can break along particular, regular planes of weakness called *cleavage planes*, seen by inspection, or by actually causing the specimen to cleave on applying a penknife blade. A *parting plane* is simply such a plane that is less well developed than a cleavage plane, whereas a *fracture* is an irregular breakage surface.

Properties such as *lustre* are described by terms such as metallic, resinous, pearly, glassy, or adamantine (diamond-like), and which are self-explanatory. A few minerals have one or more very distinctive properties which make identification straightforward. For example, the black metallic iron oxide mineral magnetite can deflect a compass needle due to its magnetic properties. One of the forms of the iron sulphide mineral pyrrhotite has similar behaviour but is brassy and metallic, and very different in appearance to magnetite. Density is rarely definitive, but the barium sulphate mineral baryte is unusually dense for a non-metallic mineral.

Moving from the field to the laboratory means that all minerals can be identified with certainty, provided they are present in sufficient quantity (although with modern analytical methods, even very small amounts can be identified). The methods described in Chapter 2 are extremely powerful for characterization that goes well beyond giving the mineral concerned a name. Compositions in terms of major, minor, and trace element content, crystal structure, or surface chemistry can all be determined in detail, as we shall see.

A note about crystal structures and compositions

The two essential characteristics of any mineral are its chemical composition and its crystal ('atomic') structure. We need to emphasize some points which are central to a proper understanding of minerals. The first concerns crystal structure, by which we mean the regular arrangement of atoms referred to above in the definition of the word 'mineral'. Many mineral 'phases' (the term used to describe a specific composition and structure) can undergo a 'phase transformation' when subjected to a change in conditions such as an increase in temperature or pressure. A phase transformation is the way in which the structure responds to changing conditions; for example, by adopting a more compact arrangement of its atoms as they are forced together at high pressure. The best-known example of a high pressure phase transformation is that involving carbon (see Figures 1b and 1c). At atmospheric pressure, carbon occurs as the mineral graphite with a structure in which layers of carbon atoms form six-membered rings with quite strong bonds between them. But between the layers, the bonds are extremely weak. At very high pressures a different structure becomes stable, that of diamond. In diamond, each carbon atom is bonded to four other carbon atoms at the corners of a tetrahedron, and each of these carbon atoms is bonded to another four carbon atoms which are themselves bonded to another four carbon atoms (and so on in all directions). This structure is an exceptionally strong arrangement and this is reflected in the fact that diamond is by far the hardest known naturally occurring substance. This is a dramatic contrast with graphite, which is one of the softest of all minerals. Whereas phase transformations at high pressures are to more compact crystal structures, transformations at high temperatures are to more open structures.

As we shall discuss in later chapters, phase transformations are of great importance for understanding our planet, given the high temperatures and pressures in the interior of the Earth. Also shown, as Figure 1d, is a synthetic form of carbon consisting of 60 carbon atoms forming a ball. This material is called *buckminsterfullerene* after the American designer and architect, R. Buckminster Fuller and because of the fancied resemblance of this molecule to his architectural invention, the geodesic dome. Another synthetic form of carbon is named *graphene* and is essentially a monolayer of carbon atoms like a single layer of atoms in the structure of graphite (see Figure 1b). These synthetic *nanomaterials* (see Chapter 4 for further discussion) have the potential for wide-ranging practical applications.

Minerals are not pure, simple chemical compounds of the kind we might find in a jar in a chemistry laboratory. Take the example of the mineral pyrite (FeS_2). We could go to our laboratory, take some powdered pure iron and pure sulphur, seal them in a glass tube from which we exclude any air, and heat this mixture up to several hundred degrees. On cooling and breaking open the glass tube, we would find a single 'phase' to be present which (on examination using methods discussed below) we could identify as pure 'synthetic' pyrite (FeS_2). Scientists, myself included, undertake such experiments either to find out the temperature

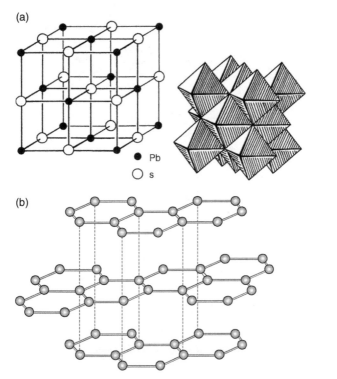

1. Crystals and crystal structures: (a) the crystal structure of galena, PbS, shown as both a 'ball and spoke' figure with black spheres representing atoms of lead and white spheres of sulphur, as well as a visualization of the structure as 'edge-sharing' PbS_6 octahedra; (b) 'ball and spoke' representation of the structure of graphite

11

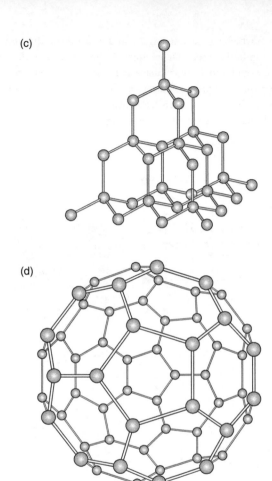

1. (c) of diamond, and (d) of buckminsterfullerene

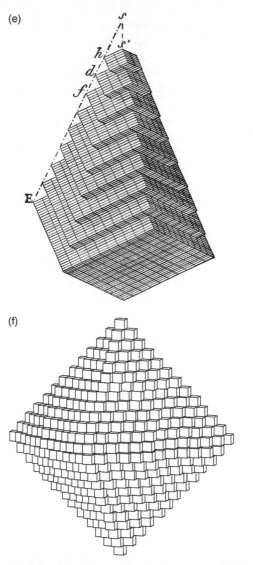

1. (e) one of Haüy's original figures showing how a crystal of calcite can be considered as built from rhombohedral units; (f) an octahedron built by stacking cubelets

13

conditions under which particular minerals are stable, or to have a sample of pure pyrite to use in other experiments.

There is an important difference between our synthetic pyrite and a natural sample of pyrite (or any other mineral) collected in the field or from a mine. The latter will always contain impurities. Some of these impurities may be minute grains of other mineral phases trapped within the pyrite, others may be held as atoms of the impurity element actually within the crystal structure of the pyrite. Although some of these may occur in vacant 'spaces' within the structure, more commonly they will be found replacing some of the iron (Fe) or sulphur (S) atoms in the pyrite. In the extreme case, a large percentage or even all of the Fe or S atoms will be replaced, as can happen with nickel (Ni) replacing the iron in pyrite to give the mineral vaesite (NiS_2). In fact, it is possible to find in nature every composition between FeS_2 and NiS_2. This is represented in mineral formulae by writing $(Fe,Ni)S_2$ and it is an example of what is known as a 'solid solution'. Many examples of this are found in nature and it is behaviour that is an especially important feature of the 'big ten' minerals, as can be seen from an example such as the olivines (($Mg,Fe)SiO_4$; see also Table 1). As well as being important for understanding the chemistry of many mineral groups, impurities may provide important clues about the way certain minerals have formed, or may be important for economic or environmental reasons. Pyrite, for example, may contain small but economically important amounts of gold (ironic given that it is sometimes called 'fool's gold'), or may contain arsenic (substituting for sulphur), presenting a hazard for human health.

Crystallography

It will be evident from the definition of 'mineral' given above that the study of minerals is also the study of crystals. Indeed, many mineralogists have also been crystallographers. Long before the discovery of X-rays and our ability to probe the ordered arrangement of atoms in crystals, scholars were attempting to understand the nature and properties of crystalline minerals. One such scholar was the Frenchman René Haüy who published his ideas in 1784. A story told about Haüy, considered by many to be the 'father of crystallography', was that whilst examining a group of crystals of calcite ($CaCO_3$) in the mineral collection belonging to a friend, he dropped the specimen which broke apart along a single plane. His forgiving friend presented him with the broken crystal which he took away and attempted to break ('cleave') in other directions. In this he succeeded, in his words, in 'extracting its rhomboid nucleus' from the crystal. This led him to propound the view that continued cleaving would eventually lead to the smallest possible unit, or building block, by repetition of which the whole crystal is built up (Figure 1e).

We now know that there is indeed a fundamental building block of any crystal that we call the *unit cell*, not obtainable in quite the way in which Haüy envisioned because it is at the atomic scale, being the smallest group of atoms from which the crystal can be built up by repetition in three dimensions. Shown in Figure 1e is one of Haüy's original figures and Figure 1f shows how a macroscopic octahedral crystal can be built up by stacking cubelets. Of course, in an actual crystal there would be billions of such cubelets. The unit cell of a crystal of galena (PbS) is a cube which can be stacked in this way to form an octahedron, or simply a cube. In Figure 1a, the origin of the cube crystal of galena can be seen in the arrangement of the lead (Pb) and sulphur (S) atoms shown in a 'ball and spoke' figure. This is one of the ways of representing the arrangements of the atoms in a crystalline solid. Also shown in this figure is the way in which each lead atom is surrounded by six sulphur atoms at the corners of an octahedron, and how these octahedra are stacked together to build up the structure. The mineral halite, sometimes known as rock-salt (NaCl), has the same crystal structure as galena. (We should note that there are some subtleties regarding the precise definition of the 'unit cell' best explored by further reading; a good account is given in the book by Putnis, details of which are given in the Further reading).

The link between the unit cell and the crystal is particularly made when we talk of *symmetry*, which is the property that distinguishes crystals from most other natural materials. It is the determination of the *elements of symmetry* of a crystal that enables the mineralogist to assign it to one of seven *crystal systems*, an important step in routine identification. The unit cells of the seven systems are illustrated in Figure 2, along with their key symmetry elements which can be planes of symmetry, axes of symmetry or a centre of symmetry. A plane of symmetry is also known as a mirror plane whereby the crystal on one side of a plane slicing through the crystal is a mirror image of that on the other side. These planes are shown by the heavier lines in the figures, which represent the intersection of the mirror plane and the surface of the crystal. (One such plane is 'shaded in' in Figure 2b). An axis of symmetry can be envisaged by thinking of holding up a large crystal between thumb and forefinger and rotating it through 360°. If, to the observer, the crystal appears the same 2, 3, 4, or 6 times during one rotation, it is said to have a two-, three-, four-, or sixfold axis of symmetry. In the figures, the symmetry axes are shown as lines passing through the centre of the crystal, and with appropriate symbols on their ends (hexagon, square, triangle, or lozenge for 6-, 4-, 3-, 2-fold axes). A centre of symmetry is present if each face has a similar parallel face on the opposite side of the crystal.

As well as the elements of symmetry, the unit cells shown in Figure 2 are characterized by the dimensions along, and angular relationships between, what we can define as their x, y, and z axes. The triclinic unit cell (Figure 2a) is the least symmetric, having only a centre of symmetry as symbolized by the open circle. Here, the dashed lines can be considered the x, y, and z axes; the lengths of the unit cell edges along these three axes are the unit cell parameters and labelled a, b, and c (where a is a length along the x axis, b along the y axis, and c along the z axis). In the triclinic case, these lengths are all different. Furthermore, none of the angles between the axes (labelled α, β, γ) are right angles. This means

16

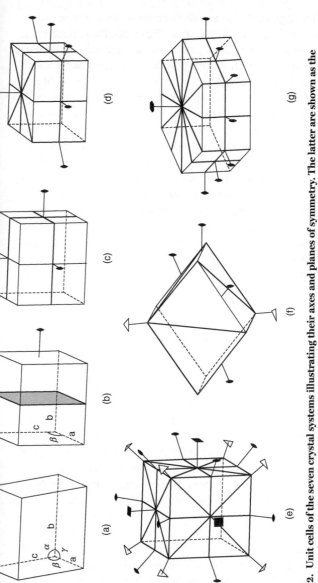

2. Unit cells of the seven crystal systems illustrating their axes and planes of symmetry. The latter are shown as the darker lines on the surfaces of the unit cells but shaded in for emphasis in the monoclinic case, (b); the axes of symmetry are the lines extending out and marked with symbols to show whether 2-, 3-, 4-, or 6-fold axes. The unit cells are: (a) triclinic; (b) monoclinic; (c) orthorhombic; (d) tetragonal; (e) cubic; (f) trigonal; and (g) hexagonal in symmetry

17

that the angles between a triclinic crystal's edges or faces are never 90°. In a monoclinic crystal (Figure 2b) the parameters a, b, and c are again unequal but, whereas the angle between the x and z axes (β) is greater than 90°, the angles between x and y, and between y and z are both 90°. Also shown here is the 2-fold axis and, at right angles to it, a plane of symmetry, a combination characteristic of monoclinic crystals. For the other five crystal systems, the angles between x, y, and z axes are always 90°. Whereas in orthorhombic crystals the a, b, and c parameters are all different, in tetragonal crystals, a and b parameters are the same and differ from the c dimension (Figures 2c, 2d). Of course, these crystals have many more planes and axes of symmetry and this reaches the extreme in the cubic system (Figure 2e) where $a = b = c$. The trigonal (or rhombohedral) and the hexagonal systems have a z axis which is, respectively, 3-fold or 6-fold and with a c dimension that differs from a and b which are themselves equal. To deal with the 3-fold and 6-fold rotation in these systems, a fourth axis in the x–y plane is introduced and labelled 'u' (so that x, y, and u are at 120°).

The seven crystal systems are just the beginning as regards crystallography, one of the oldest formal branches of science. Subdivision of the seven crystal systems on the basis of symmetry elements gives us the 32 crystal classes. Here, whereas each system has a minimum requirement in terms of symmetry needed to 'qualify for membership', other classes within that system have additional symmetry elements, but not sufficient to qualify for membership of another system. For example, in the cubic system, the 4-fold axes seen in the cube itself (Figure 2e) are not essential for membership of the system. The essential elements are the four 3-fold axes that, in the cube, pass through the opposite 'corners' of the crystal. Therefore, a lower symmetry crystal belonging to the cubic system can be in the form of a tetrahedron, which has four triangular faces giving a shape familiar from some packaging of milk or other drinks. Also, as discussed below, the basic building block of all silicate minerals is the SiO_4 tetrahedron (see Figure 3a).

One other point worth emphasizing is that a particular crystal shape may be found in several different systems. A simple example is provided by the octahedron, already illustrated for the cubic case in Figure 1f. If the same shape is modified by 'stretching' along the z axis, lengthening the c dimension so that it is now greater than the a and b dimensions, the octahedron becomes a tetragonal rather than a cubic octahedron.

(a)

(b)

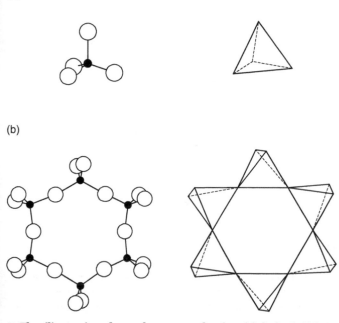

3. The silicate mineral crystal structures showing: (a) the basic SiO_4 building block as both 'ball and spoke' model and as a tetrahedron; (b) a six-membered ring of linked tetrahedra

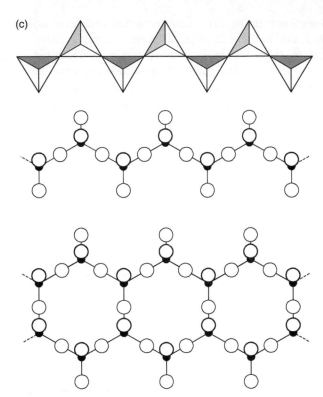

3. (c) a chain and double chain of linked tetrahedra

Mineral families

An important aspect of the development of mineralogy, as with other natural sciences, has been concerned with classification. Minerals are classified in terms of membership of families, better termed *mineral groups*. The main groups are shown in Table 1. As with many classification schemes, there are various ways we could choose to organize the ~4,400 known mineral *species* into groups. We could, for example, regard all of the copper-containing

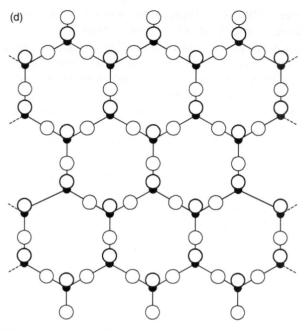

3. (d) a complete sheet of linked tetrahedra

minerals as members of one group. This would involve naturally occurring copper metal itself (*'native'* copper), copper combined with *anions* (i.e. negatively charged atoms) of elements such as sulphur, oxygen, chlorine as *sulphides*, *oxides*, and *chlorides* of copper. There would also be copper in combination with more complex anions such as sulphate (SO_4), carbonate (CO_3), and silicate (SiO_4). However, both in terms of mineral chemistry and natural occurrence, it is more useful to classify minerals such that all of the sulphides form a group, and likewise all of the oxides, carbonates, silicates, and so on.

In Table 1, only the main groups are shown with some important examples for each group. It is common practice to include closely related minerals within a major group; for example, including hydroxides (OH containing minerals) in with the oxides, or including chemically related minor minerals within a group (selenides and tellurides within the sulphide group). The silicates are a special case when it comes to classification, given their overwhelming dominance of the 'top ten' minerals which is a consequence, in turn, of their dominance in the rocks of the Earth's crust and upper mantle. All silicates contain tetrahedral building blocks in which a silicon atom is located at the centre of four oxygen atoms, thus forming a SiO_4 unit (see Figure 3a). These units can be represented by a 'ball and spoke' figure, or simply as a tetrahedron if the intention is to show how they link together to build up the structures of the different silicates. Tetrahedra can remain unconnected as 'islands' held together by the other atoms in the crystal; in the olivines these are atoms of magnesium and/ or iron. There may be pairs of tetrahedra joined by sharing a single oxygen, rings with different numbers of units linking together, or single chains as in the pyroxene minerals (see Figures 3b, 3c). The amphibole family has double chains, and the micas and clay minerals complete sheets of linked tetrahedra (Figures 3c, 3d). Linkage via the sharing of all four corner oxygen atoms produces the framework structure of quartz. In some cases, AlO_4 tetrahedral units may also be involved in building up the structures instead of SiO_4 units; this is the case for the framework structures of the feldspar minerals, the most abundant of all minerals in the rocks of the Earth's crust. The most important silicate minerals and their categorization on the basis of structure are shown in Table 1.

In our discussion of minerals, I will focus on those that are either the most abundant, as noted above in terms of the 'top ten minerals', or of particular interest as regards mineral resources or Earth processes (such as those associated with the deep interior). The great majority of minerals, and even whole groups of minerals, are relatively rare and will not feature in our discussions. Nevertheless, the principles involved in the methods of study, and the processes of formation of these rarer minerals are essentially the same as for many of the commoner minerals which are discussed in this book. There are comprehensive mineral classification schemes, as well as listings of most known minerals, detailed in the recommended further reading at the end of this book.

Chapter 2
Studying minerals

The study of minerals begins with their characterization, identification, and classification. Historically this has been based on properties observable in hand specimens such as crystal morphology (revealing crystallographic properties such as symmetry elements), cleavage, hardness, density, lustre, streak, and, in some cases, colour. For a small number of minerals, other properties may be diagnostic, such as magnetic behaviour. These properties were used as the basis of the first serious attempts at classification, dating back as far as the physician–natural historian Georgius Agricola who, in his work *De Natura Fossilium* published in 1546, classified about 600 minerals. With the developments in chemistry in the 19th century, it became possible to determine the chemical compositions of minerals and to put classification on a firm footing.

Shining light on minerals

A major advance in studies of both minerals and rocks is associated with the optical microscope. Although single lens magnifying glasses were mentioned by roman writers in the 1st century AD, the first multiple lens device was invented around 1590 by the Dutch spectacle makers Zaccharias and Hans Janssen. It was later improved upon by many inventors including Galileo who, in 1609, worked out the optical principles involved as well as improving the design. Using such instruments, mineral grains or fragments could be examined in the same way as other naturalists might examine a plant fragment or a small insect. A very important advance occurred in 1815, associated with the work of the Scottish scientist William Nicol. It was found that by preparing a very thin slice (30 micrometres = 0.3 mm thick) of a mineral or rock and gluing it to a glass slide, for most minerals, light would pass through (be 'transmitted' through to use the correct terminology) the resulting *thin section*. It was also found that by passing the incident light through a polarizer, similar to the 'polaroid' material used in sunglasses, various new optical properties of the minerals studied in *plane polarized light* could be observed which aided in their identification. If, in addition to polarizing the light before it passes through the mineral, the light has to pass through a second polarizing filter but this one at an angle of 90° to the first (using so-called observation under *crossed polars*), a further range of diagnostic optical properties could be observed. These and other related developments concerning the ways in which light passes through the minerals in a thin section, revolutionized the study of minerals and rocks because the mineral grains could be readily identified even when very small.

Furthermore, in rocks the shapes and sizes of individual mineral grains and their interrelationships could be studied, providing insights into how various rocks have formed or have been modified since their formation. A photograph of a typical thin section of an igneous rock viewed under the polarizing microscope is shown in Figure 4a. Here, a thin section of a rock from the island of Rhum in north-west Scotland has been photographed under crossed polars. This section shows a rock which has crystallized from a melt. Two minerals are present; the first is the 'island' silicate mineral, olivine (see Table 1). All of the rounded and 'crystal-shaped' grains of different colours (shades of grey in this black-and-white photograph) are olivine crystals in different orientations (which is the reason for their different colours). The other mineral is the 'framework' silicate mineral plagioclase feldspar; this forms the matrix in which the olivine crystals are set. The plagioclase is of two different orientations, with that above a line from the bottom left-hand corner to roughly midway along the right-hand side appearing white, and below that line appearing grey. The olivine must have crystallized first when this rock was formed, followed by the plagioclase feldspar. Note the white bar in the bottom right of this photograph is a 1 mm scale bar.

A limitation of the transmitted light microscope is that the minerals being studied need to be translucent. The great majority of the minerals that make up common rocks, such as all of the silicates and carbonates, do transmit light through a 30 micron thin section. However, most metal sulphide and metal oxide minerals do not; that is to say, they are *opaque*. Here, the related technique of reflected light microscopy can be used. In this case, a roughly 1–2 cm diameter piece sawn off the sample is mounted in cold-setting plastic resin and a flat surface is ground and polished to give a mirror-like surface. This *polished section* is studied using a microscope where the light (again, plane polarized) is reflected back to the observer from the polished surface via various magnifying lenses. Observations can again be made under crossed polars, and various optical properties used to identify the minerals, although identification is often not as straightforward as in transmitted light and ancillary measurements (such as of the percentage of the incident light reflected back from the polished surface) may be needed.

As with transmitted light observations, a very valuable aspect of reflected light microscopy is the information that can be provided on the sizes, shapes, textures, and interrelationships of the minerals studied. A photograph taken under the microscope of a polished section in plane polarized light is shown in Figure 4b. The sample here comes from deposits mined for nickel and located at Sudbury, Ontario, Canada. In this case the rounded grey grains are of the iron oxide mineral magnetite, the dominant

4. Minerals seen using the optical microscope: (a) a thin section photographed in transmitted light and under crossed polars of a rock with olivine crystals in a matrix of plagioclase feldspar

(b)

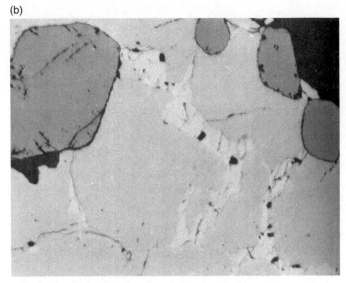

4. (b) a polished section photographed in plane polarized reflected light of an ore sample containing pyrrhotite, pentlandite, and magnetite

mineral present is the iron sulphide pyrrhotite; this forms interlocking grains between which there are what appear like 'veinlets' of the mineral pentlandite ($(Ni,Fe)_9S_8$) which is the source of the nickel. The width of this field of view is 1.2 mm. (Note that the black regions are pits in the polished surface.) The minerals in this sample have crystallized from a predominantly sulphide melt, with the magnetite forming first and then an iron–nickel sulphide phase. On cooling, most of the nickel has diffused to grain boundaries and formed the pentlandite. Because the reflected light technique is mostly used to study ore minerals, it is also known as *ore microscopy*. It is also possible to combine the advantages of thin and polished sections by preparing a thin section and then polishing its surface. The sample preparation methods needed in order to study minerals under the optical microscope are also required for certain more advanced analytical methods, such as the electron microprobe which is discussed below.

X-ray vision

The developments in optical microscopy are part of a story that began in the 16th century and could perhaps have been foreseen. Those associated with a great late 19th century scientific advance were wholly unexpected. X-rays were accidentally discovered in 1895 by the German physicist Roentgen, leading to their familiar use in medical diagnosis. However, it was not until 1912 that von Laue and his co-workers demonstrated that X-rays could not only pass through a crystalline solid such as a mineral, but also produce a characteristic pattern on a photographic plate, a pattern that can be used to determine the arrangement of atoms in the solid, i.e. its 'crystal structure' (see Figures 1a, 1b, 1c; also shown in Figure 5a is the most famous diffraction pattern ever recorded, that of a single crystal of DNA). The great breakthrough in this practical application of X-rays came with the work of the father and son team of William and Lawrence Bragg. Bragg's Law established the mathematical relationship between the wavelength (energy) of X-rays passing through the solid and the distance separating the planes of atoms in the crystalline solid, opening up the whole field of understanding of such solids at the level of individual atoms, and how their ordered arrangements are reflected in the external morphology of crystals. As illustrated in Figure 5b, any crystal can be considered as made up of numerous planes of atoms, with adjacent planes separated by a distance d. It happens that this distance (the 'd spacing') is of the same order of magnitude as the wavelength of the X-rays that can be generated in the laboratory and symbolized by λ. This means that when a particular plane of atoms is at the so-called Bragg angle (θ), the X-ray beam will be 'reflected' and this X-ray reflection can be detected as spots on a photographic film. So, in the Bragg equation which can be written as '$\lambda = 2d \sin \theta$', the value of λ is known or can be controlled, and of θ can be measured in an experiment to give the value of d. Although this process is not really one of 'reflection' but rather of 'diffraction', to simplify matters in this discussion it can be thought of as reflection.

When all the d-spacings between all the planes in a given crystal have been measured, and data also collected on the intensities (brightness) of reflections which provide further information on the elemental identities of the atoms in the structure, then the crystal structure of the mineral, or other crystalline solid being investigated, can be calculated.

Amongst the first crystal structures determined were the relatively simple structures of minerals such as halite (NaCl, rock-salt) and the sulphides of lead (PbS, galena; see Figure 1a), zinc (ZnS, sphalerite), and iron (FeS_2, pyrite). Much of this work was completed by Lawrence Bragg when he was Professor of Physics at Manchester University. The work of Bragg on the crystal structures of minerals laid the foundations of modern X-ray crystallography. The last major rock-forming silicate mineral structure to be determined was that of feldspar, solved by W. H. Taylor in 1932 (on Christmas Day!). These early years saw painstaking measurements of the relative positions and intensities ('brightness') of spots recorded on photographic film in order to elucidate crystal structures. Many months of work were needed to determine more complex structures; in those days, doctoral degrees were commonly awarded for the 'solving' of one or two such structures. Today, photographic methods are no longer used; the data are recorded using computer-controlled diffractometers in which X-ray detectors scan the space around the crystal to determine at which angles X-rays from the incident beam have been diffracted and their intensities. The data acquired in this way are analysed and interpreted using standard software. The solving of crystal structures that would have taken many months in the early years can now be completed in a matter of days.

Crystal structure determination requires a single crystal (such as a single 'cube' of halite or galena) with a diameter of a millimetre or so. Such a crystal can be mounted (glued) on to the end of a thin glass fibre and then placed in the path of the

X-ray beam in a camera or, in modern equipment, a diffractometer. It is also possible to use the data collected when the X-ray beam is directed on to a finely powdered sample of a mineral (or mixture of minerals) spread on a glass slide. In this case, peaks in X-ray intensity are recorded as the detector scans through a range of angles. Again, using Bragg's Law, these peaks can be converted into spacings between layers of atoms in the structure (*interplanar spacings*) which are characteristic for the structure and in most cases a 'fingerprint' means of identifying the mineral or minerals present in the powder. This method is called *x-ray powder diffractometry* and is the most widely used method for identifying minerals or other crystalline solids in powder form. Typical data from an X-ray powder diffraction experiment are shown in Figure 5c. As the diffractometer scans through a range of values of θ (measured in degrees), reflection

(a)

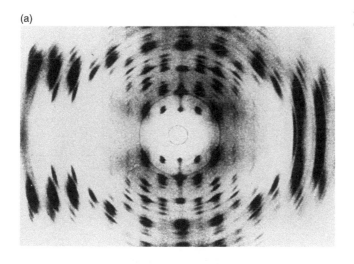

5. X-ray diffraction: (a) a diffraction pattern on a photographic film from a single crystal of DNA

(b)

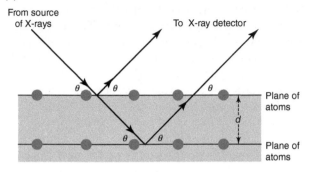

(c)

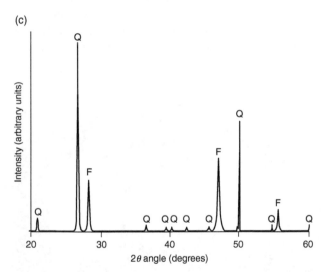

5. (b) Bragg diffraction from layers of atoms in a crystal; (c) a typical X-ray powder diffraction pattern for a mixture of quartz (Q) and fluorite (F)

of X-rays produces peaks in the diffraction pattern at values characteristic of the mineral. If there is more than one mineral present, the peak intensities are proportional to the amounts of those minerals present.

The power of the synchrotron

The radiation employed in an X-ray diffraction experiment is of a single energy (or wavelength). It is electromagnetic energy which, in one of the great paradoxes of modern physics, can be described both as waves or as a stream of particles ('photons'). It is fundamentally the same as the light we detect with our eyes, except that the latter is of much lower energy (longer wavelength). In fact, electromagnetic energy extends as a continuum from the very long wavelengths and low energies of radio waves, through the infrared, then visible light, then the ultraviolet, and on to the high energies of X-rays and, ultimately, gamma-rays. Electromagnetic radiation in nearly all of these ranges can be used in experiments to probe the nature of materials including minerals. Earlier experiments used what are now called 'conventional' sources of radiation, not very different from the domestic light bulb, and certainly on a 'bench-top' scale.

However, a major advance took place with the availability of *synchrotron radiation*, beginning in the 1950s. At a synchrotron, radiation is generated when a linear particle accelerator 'fires' electrons at near light speed into a 'storage ring' which is several hundred metres in diameter (see Figure 6a). As the electrons travel around the ring, kept on a curved path by very large magnets and other devices, they emit electromagnetic radiation (photons) at a tangent to the direction of travel of the electrons. This radiation can range in energy from infrared through visible light to the ultraviolet, and low energy ('soft') X-rays to high energy ('hard') X-rays. The great advantage of synchrotron radiation is its intensity, which is orders of magnitude greater than that from conventional laboratory sources. Also, it is so-called 'white' radiation which, as with 'white light' (such as daylight), contains contributions from a range of energies in a single beam (as in the rainbow colours that make up white light). This means that in an experiment, a specific energy can be selected to interact with the sample, or a range of energies can be scanned.

The X-rays provided by a synchrotron can be used in diffraction experiments in the same way as those from a conventional laboratory X-ray source. But, as already emphasized, the advantage of synchrotron radiation is its much greater intensity. Using modern detectors, not only can the data needed for determining crystal structures be easily acquired, it can be acquired very rapidly (sometimes in fractions of a second). This means that changes in crystal structure in response to changing conditions can be studied as they occur (in 'real' time). At the synchrotron, diffraction data can also be obtained from powder samples and of sufficient quality to be used to determine crystal structures; this is not the case for data obtained using a conventional source. At a synchrotron, experiments can be successfully performed on samples with very low concentrations of the substance of interest, or on very small quantities of sample. The synchrotron beam can also be focused down to a small spot of diameter about one micron or even less, so as to study individual particles, grains, or particular regions of interest in the sample.

Many different types of experiments can be performed at the synchrotron, including the scattering of X-rays from particles suspended in a fluid which can tell us about mechanisms of crystallization and growth from solution. However, along with diffraction, the most important measurements have generally been those involving the absorption of X-rays (or ultraviolet photons) by a powder sample or a slurry. The subject of synchrotron methods is one that is awash with acronyms, most of which need not concern us here. However, two of particular importance are EXAFS (extended X-ray absorption fine structure) and XANES (X-ray absorption near edge structure) spectroscopies. In the simplest EXAFS experiments, an X-ray beam emerging from the synchrotron passes through a detector that measures its intensity, and then through the sample before entering another detector which can determine if X-rays have been absorbed by the sample. Between the emerging beam and the first detector is a device called a *monochromator* (meaning literally 'one colour' and a reference to selecting a single colour from a beam of white light which, in this case, is not visible light but 'white' X-radiation). This device is used to scan through a range of X-ray energies. At certain energies, X-rays are absorbed by the sample, as seen from the differences in intensities recorded by the detectors placed 'before' and 'after' the sample. Why are X-rays absorbed by the sample? This occurs when electrons that 'orbit' the nucleus of the atom of a particular element of interest (say iron, for example) are 'kicked' (or more correctly *excited*) into a higher energy state by the energy provided by the X-rays. This can only happen at a particular energy that is characteristic of the element concerned, and of the electrons at a particular distance from the nucleus (in what is termed a particular *electron shell*). What is seen in the experiment, as the monochromator is scanned through a range of energies, is an energy at which the absorption of X-rays dramatically increases; this is an *absorption edge* (see Figure 6b).

(a)

6. **Synchrotron radiation showing: (a) an aerial view of the European Synchrotron Radiation Facility**

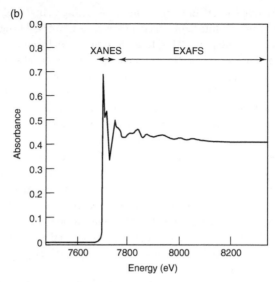

(b)

6. (b) a typical X-ray absorption spectrum showing the absorption edge and EXAFS and XANES regions

The presence of the absorption edge at a particular energy tells us that the element concerned is present, but that is not the main reason for interest in EXAFS and XANES spectra. If the element being studied were to be the only element present, and to be in the form of a gas rather than a solid or liquid, then the spectrum (a plot of incident energy against X-ray absorbance) would be a smooth line, rising dramatically at the absorption edge, reaching a maximum and then gently falling away. However, if the element of interest is in a solid or liquid form, on a surface, and especially if it is associated with atoms of other elements, the spectrum will exhibit *fine structure*. This is the name given to smaller peaks and 'wiggles' seen near the absorption edge (the XANES) and in the region beyond the edge (the EXAFS; see Figure 6b). The interpretation of these features, achieved using computer fitting and modelling, provides key information about the nature and environment of

36

the element concerned in the sample under study. Suppose, for example, we are studying iron in a mineral system. The XANES and EXAFS spectra can tell us whether the iron is present as metallic iron, as ferrous or ferric iron, and if it is bonded to other elements, then what it is bonded to (such as oxygen or sulphur) and how many atoms it is bonded to (most commonly four or six) and at what distances. The atoms immediately surrounding the iron in the so-called 'first shell' can always be detected and, in many cases, further shells of atoms at greater distances can be pinpointed and their distances away measured. In this way the structure of a solid mineral, even one that is poorly crystalline, can be further defined, or the environment of the iron in a solution or at a surface characterized.

The probing electron

The various forms of electromagnetic radiation are powerful probes of the nature of materials including minerals. Much can be learnt from studying the ways in which a beam of electromagnetic radiation is diffracted, absorbed, or reflected, or causes the emission of other forms of radiation. But samples can also be bombarded with particles and the resulting interactions or emissions studied. The particles used include protons or various ions such as those of oxygen or caesium but the most important, by far, are electrons. Electrons are used in the electron probe microanalyser and in various types of electron microscope.

In the electron probe, a beam of electrons is focused down to a very small diameter using magnetic lenses. The solid mineral sample is prepared as a polished section or polished thin section as described above. When the electrons strike the flat polished surface of the sample, they 'knock out' electrons from inner electron shells of the atoms present, causing electrons from outer shells to drop down and fill the 'holes' created by this process. The energy lost when this occurs is emitted as X-rays of an energy that is characteristic of the atom concerned. In the electron probe, this radiation is detected, showing the presence of atoms of the element concerned. As well as the energy indicating the presence of the element, the intensity of the emitted X-rays measured against a standard of known composition provides the means to a quantitative analysis of the sample.

Since its development in the 1950s, particularly through the work of Castaing in Paris, it is no exaggeration to say that the electron probe has revolutionized the study of minerals. This is because a volume of mineral as small as a few cubic micrometres (microns) can be analysed. In samples prepared as polished thin sections or polished blocks, individual grains or small areas of a sample can be quantitatively analysed following examination under the optical microscope. By rapidly scanning the beam over an area of the polished surface, it is also possible to produce 'maps' of the distribution of major, minor, or trace elements between different areas of the sample. All but the very lightest elements (H, He, Li) can be analysed in concentrations from roughly 0.5 to 100 per cent and in favourable cases, much lower concentrations can be analysed (approaching tens of parts per million). In the 60 years since the first electron probes became commercially available, this instrument has been essential for the discovery of hundreds of new minerals, many of which occur only as small grains first seen under the optical microscope.

If the electron probe has revolutionized mineralogy, the various types of electron microscopes have also had a tremendous impact on the characterization of minerals. In the traditional instrument, the transmission electron microscope (TEM), an electron beam is focused down so as to strike the sample as a very small spot. In this case, however, the specimen must be very thin so that the electron beam can pass through it, producing an image on a fluorescent screen. An appropriately thin specimen can be produced by boring a very small hole in a thin slice of mineral or rock using a beam of ions. There are regions around the edges of the hole, only a few tens of micrometres in extent, which are generally thin enough to allow passage of the electron beam. Alternatively, areas at the edges of very small grains may be suitable, the small grains being produced by crushing of a larger sample or being the natural state of the sample (as for many precipitates; see Figure 7a, where the grain sizes and shapes of a synthetic iron oxide are evident, with a magnified insert showing the fine structure at the edges of some grains). Recent decades have seen further refinements in the preparation of samples for study using the TEM. In particular, samples for putting into the microscope which are too small to be seen with the naked eye can be prepared by slicing through the sample using a focused ion beam (FIB).

The spatial resolution of the TEM, particularly that achievable using a variant of the technique known as High Resolution Transmission Electron Microscopy (HRTEM), is impressive (0.2–0.5 nanometres (nm)). It is sufficient to image features at the level of unit cells (which, for simpler structures, generally have dimensions of the order of 0.5–2.0 nm) if not to image actual atoms. An HRTEM image of the serpentine mineral chrysotile is shown in Figure 7c and clearly illustrates our capability to image at the level of unit cells, making the link between the abstract world of crystallography and our direct observations. In the image of this fibrous (asbestiform) mineral, we are looking down the fibre axis, and the smallest features (dimensions about 0.5 nm) are the unit cells in this structure.

Two other features of the TEM add to its power for studying minerals and many other materials. The first is that the electrons striking the sample can also be diffracted like X-rays, producing a pattern of spots as in an X-ray diffraction experiment. The positions and intensities of these spots provide structural information and a 'fingerprint' identification of the material being investigated. For example, the data in Figure 7b confirms

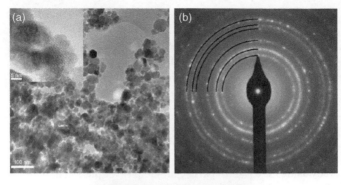

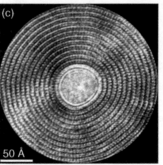

7. Transmission electron microscopy (TEM) data obtained from nanometre-sized magnetite particles with: (a) an image showing grain sizes and shapes (white scale bar is 100 nm) with an inset image at higher magnification (scale bar is 5 nm) showing details of structure; (b) a selected area electron diffraction pattern confirming that the grains are of magnetite; (c) an HRTEM image of the layer silicate mineral chrysotile ('asbestos'); scale bar is 50 ångstroms (Å)

that this material is the iron oxide mineral, magnetite. The second feature available on many modern instruments is a capability for using emitted X-rays in chemical analysis. In this case, the X-rays emitted from the sample when struck by the electron beam can be used to obtain a chemical analysis, rather as in the electron probe, although such analyses generally provide only approximate values.

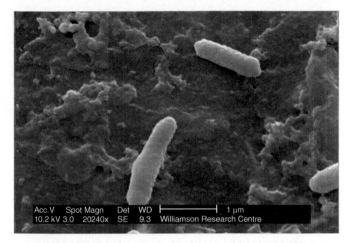

8. Environmental scanning electron microscope (ESEM) image showing several 'grub-like' bacteria on an iron oxide mineral substrate

Another type of electron microscope widely employed in the study of minerals is the scanning electron microscope (SEM). As the name suggests, in this instrument the electron beam scans very rapidly across the sample surface. Secondary electrons are emitted from the surface and these are used to form an image of the sample even if the surface is rough, giving a three-dimensional effect (see Figure 8). Although the resolution is not as great as for TEM (generally it is in the range 10–20 nm) samples can be put in the instrument without special preparation. The X-rays emitted from the bombarded sample can again be used to obtain a semi-quantitative chemical analysis. A disadvantage of the conventional SEM and also of TEM and the electron probe is the need to work with the sample in a vacuum. In recent years, technical advances have led to the development of the 'environmental (E)SEM', an instrument where samples which are hydrated or coated with delicate biological materials such as biofilms or microbes, can be studied without being damaged. This is especially important in environmental work. The image shown in Figure 8 is actually from an ESEM and shows an iron-reducing bacterium (*Geobacter sulfurreducens*) on an iron oxide substrate.

Seeing atoms with scanning probe microscopes

The SEM is primarily an instrument for obtaining a much magnified image of the surface of a sample. Surfaces are a particularly important aspect of modern studies of mineral systems. This is because change generally takes place at surfaces and interfaces, as we will see in later chapters. Clearly the electron microscope methods can provide information about surfaces and interfaces but with certain limitations. Fundamental questions about how the nature of mineral surfaces at atomic level may differ from a simple termination of the bulk crystal structure, how minerals dissolve, or how they may sorb material from solution, could not be properly addressed until another dramatic technique development took place in the 1990s. This was the development of the scanning tunnelling microscope (STM) and the related technique of the atomic force microscope (AFM). Both of these methods are capable of imaging surfaces at 'atomic resolution', that is to say, at the level of magnification where individual atoms can be 'seen'.

These two methods are remarkably simple 'bench top' experiments. In the STM, a very sharp tip (like a stylus on an old-fashioned gramophone) is scanned just above the surface of a sample which must be a conducting material. At the same time an electrical potential is applied and a current flows between the tip and the sample. Fluctuations in this current are monitored to build up an image of the surface. In AFM, a sharp tip is mounted on the end of a springy arm (cantilever), and the movement of the cantilever in response to the very small forces between the mineral surface and the tip is monitored during scanning to build up the image. Both methods depend on the ability of the instrument to scan back and forth over nanometre distances; this is achieved using a 'piezoelectric' device in which applying a current across a crystal causes it to expand or contract by a very small amount (see Box 2). Both STM and AFM can operate in vacuum, in air, or with the surface in contact with a fluid. Figure 9 shows atomic resolution STM images of two different iron

Box 2 From quartz clock to crystal set

The piezoelectric effect which forms the basis of the STM and AFM methods was first discovered in quartz crystals by the Curie brothers in 1880. In this phenomenon, when certain crystals are compressed along particular directions they acquire an electric field, with one surface of the crystal becoming positively charged and the opposite surface negatively charged. This property has been widely used in technology, in devices ranging from pressure gauges to the crystal pick-ups used for record players. The converse also applies, such that when an electric field is applied to a piezoelectric crystal, its dimensions change very slightly. For example, in a piece ('bar') of crystal quartz subjected to a moderate electric field (10,000 volts/m), the change would be equivalent to a bar that is one metre long becoming 1.0000000225 metres long. This phenomenon is at the heart of the scanning probe microscope methods. By using a cleverly designed array of piezoelectric devices to control the movement of a scanning tip when a potential is applied, it is possible accurately to scan over nanometre ranges and, therefore, to image surfaces at 'atomic resolution' (Figures 9a, 9b). A more familiar application of the piezoelectric effect in quartz is in the quartz clock (or watch). In this case, a quartz crystal is made to oscillate with a very precise frequency by an electric signal. This oscillation, rather than the swing of a pendulum in a traditional grandfather clock, is used to keep time to far greater accuracy than in a mechanical clock.

Piezoelectricity is just one of a number of examples where new technologies have been developed using a property first observed in a mineral. Another example concerns the early development of radio. At the beginning of the 20th century, researchers discovered that certain semiconducting minerals could be used to detect radio signals. In a very simple radio receiver called a 'crystal set', a small piece of galena (PbS) or of pyrite (FeS_2) contacted by a fine wire (known as a 'cat's whisker') is central to the receiving device. In the early days of radio, such devices could be made at

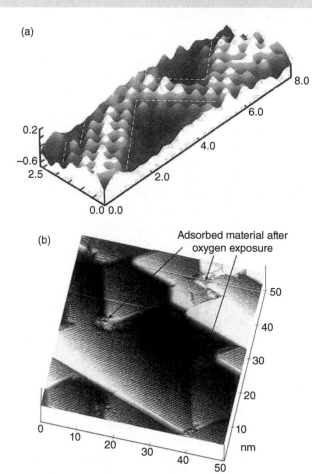

9. Scanning tunnelling microscope images of: (a) the surface of a crystal of pyrite (FeS_2) and (b) the surface of a crystal of pyrrhotite (Fe_7S_8)

Minerals

sulphide mineral surfaces. In Figure 9a, the surface is of pyrite (FeS_2) with darkened regions due to partial oxidation of the surface. In Figure 9b the surface is of pyrrhotite (Fe_7S_8) which also shows changes associated with oxidation. Note the nanometre scale of these observations.

The above account provides just a small sample of the numerous methods now available to study minerals. In my view, we too often take for granted our abilities to study the natural world, in particular the mineral world, at a level of magnification that can resolve the contributions of individual atoms. In this context, it is sobering to recall that an atom is as many times smaller compared to a soccer football as a football is compared to planet Earth. Armed with the techniques described above, and many other remarkable methods either available now or under development, we can study all aspects of minerals, their structures, chemistries, surface chemistries, and reactivities.

Chapter 3
Minerals and the interior of the Earth

The Earth is a dynamic planet, particularly when considered on the geological timescale of many millions of years. The short-term and visible evidence of this dynamism is volcanoes and earthquakes, the long-term evidence is the creation of oceans and the collision of continents. It was not until the 1950s that a theory to adequately explain this drama was propounded. The key role played by mineral magnetism in this story (as well as the role of the magnetic properties of minerals in human history) is outlined in Box 3.

Today the paradigm of plate tectonics is so widely accepted that its terminology has entered everyday speech. It is not my intention to devote space to reviewing the theory of plate tectonics; numerous eloquent accounts already cover this topic. However, the tectonic plates that are created and destroyed at different types of plate boundaries are made up of rocks with their constituent minerals, so the story of plate tectonics is also the story of the formation, transformation, and destruction of minerals. In particular, it is the story of crystallization of minerals from the molten state, and the destruction of minerals by melting. Figure 10, in a schematic and simplified cross-section, shows melts rising up from deep in the Earth along mid-ocean ridges, solidifying and forming new ocean crust, spreading apart at this *constructive* plate boundary. By contrast, the plate is dragged down (*subducted*) at a *destructive*

Box 3 From lodestone to spreading ocean floors

The discovery that the iron oxide mineral magnetite (Fe_3O_4) once called 'lodestone' (where lode means 'way' or 'course') has the distinctive properties we now associate with a particular form of magnetism was another of the great landmarks in the journey of mankind to the modern age. The ability of magnetite to attract iron was known to the ancient Greeks by around 600 BC. However, it is not until the period 1000–1200 AD that there are records of its use as a compass for navigation, helping thereafter to open the door to global exploration and to many important discoveries. The same mineral, along with a small number of other minerals with similar magnetic properties, also played a key role in a scientific revolution a thousand years later, in the 1960s.

After the Second World War, mapping of the ocean floor and its magnetic properties arising from the fossil magnetism (palaeomagnetism) frozen into the basalt rocks when they solidified on emergence at a mid-ocean ridge, provided the first evidence that the sea floor is spreading outwards on either side of the ridge. These rocks show evidence of having undergone periodic reversals in the polarity of the magnetic field in which they formed. The resulting magnetic striping of the ocean floor provided the first clear evidence for the opening of oceans, demonstrating continental drift and leading to the theory of plate tectonics. Palaeomagnetic studies, which depend on the properties of a small number of (mostly iron-containing oxide) minerals, have played a central role in unravelling the extraordinary story of opening and closing oceans and drifting continents through geological time.

plate boundary and melting occurs to form magmas which, being less dense than the surrounding rocks, rise towards the surface. If the magma reaches the surface we have volcanoes, and the magma erupts as lava or ash, but it may stop short of the surface

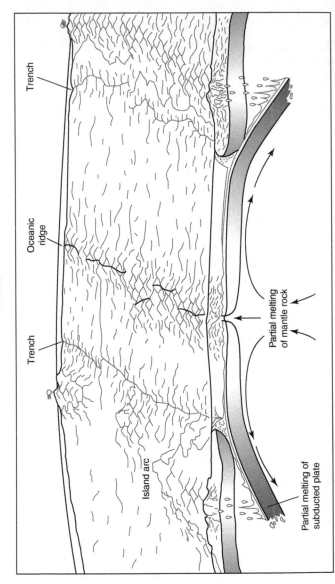

10. A much simplified cross-section through a part of the lithosphere

and crystallize slowly, forming a body of rock termed a *pluton*. Magmas may also originate at *hot spots* associated with mantle 'plumes' rising up beneath either oceanic or continental crust. Whatever the sources of magmas, mineralogy meets plate tectonics when we consider melting and crystallization processes.

Crystallization and melting

In the 1920s, an experimental scientist in the USA named Norman Bowen performed a series of classic experiments simulating the crystallization of rocks from magmas. Like others before him, Bowen found that the process of crystallization from a melt occurs over a large range of temperatures (hundreds of degrees) and that the first crystals to form have a very different chemical composition to the overall composition of the melt. As cooling continues, these crystals react with the remaining melt in quite a complicated way during the course of crystallization. For example, the first minerals to crystallize from the sort of magma that erupts at mid-ocean ridges (*basaltic magma*) are olivines (see Table 1). The olivines range in composition from Mg_2SiO_4 (forsterite) to Fe_2SiO_4 (fayalite) in a continuous series. The first to crystallize are magnesium (Mg) rich and they react with the melt, becoming more iron (Fe) rich as crystallization proceeds. However, with continued cooling and crystallization, these olivines are replaced, either completely or in part, through their reacting with the melt to form a new mineral, an Mg-rich pyroxene $(MgSiO_3)$ which again reacts to become more Fe rich and can be itself replaced by an amphibole and then by a mica (biotite). Important here is the idea of incomplete reaction, because removal of the earlier formed crystals (say by settling to the bottom of a body of magma) means effectively starting with a new melt composition. This process of *fractional crystallization* greatly increases the range of minerals and hence types of rocks that can be produced from the starting melt.

Largely overlapping with the crystallization of Mg–Fe silicates from a basaltic magma, there is also crystallization of feldspars, beginning with calcium-rich species ($CaAl_2Si_2O_8$) and with continuous substitution of calcium by sodium, and of part of the aluminium by silicon to reach the composition $NaAlSi_3O_8$. With the feldspars, the same crystal structure is retained throughout in a process described by Norman Bowen as a *continuous reaction series* and contrasting with the *discontinuous reaction series* in which the species of mineral changes as well as its chemistry (olivine→pyroxene→amphibole→mica and ultimately →potassium feldspar and quartz). What we see here is the way in which a magma can 'evolve' via the process of fractional crystallization with the changes now known as *Bowen's Reaction Series*. Indeed, Bowen entitled his classic book describing his work *The Evolution of the Igneous Rocks*.

Although Bowen suggested that all igneous rocks could be derived from a parent basaltic magma (the most common magma) by fractional crystallization, we now know that such extreme 'differentiation' is uncommon. Even so, fractional crystallization is extremely important in understanding how minerals crystallize from melts, and in explaining many phenomena observed in minerals. One of these is *zoning* in crystals such as those of olivine or feldspar (particularly in certain plagioclase feldspars formed at comparatively low temperatures). A zoned crystal exhibits 'onion-like' layers between the core and the rim. For example, the innermost core of a crystal of feldspar (the part which must have crystallized first) will commonly have a calcium-rich composition and successive layers (zones) have compositions that are more sodium rich. This is because the earlier formed feldspar crystals have not reacted completely with the melt before crystallization of later (more sodium-rich) feldspar. The minerals of the discontinuous reaction series, from olivine through to quartz, involve an increase in silica content from first formed to last formed. Also, this series shows successive changes in silicate crystal structure from island (olivine) to single chain (pyroxene) to double chain (amphibole) to layer (mica) to framework (quartz) forms.

Many of the concepts applicable when we talk about crystallization of magmas are important when we consider melting of rocks; in a sense they apply in reverse. So, just as crystallization takes place over a wide range of temperature, melting also takes place over hundreds of degrees. When melting begins, it is only a small fraction of the rock which becomes liquid, that fraction which would be last to crystallize if we were cooling instead of heating. Because only part of the rock melts, the process is called by the obvious name of *partial melting*. This is a very important process because it plays a major role in the generation of magmas as part of the plate tectonic cycle. Much of this magma generation arises from partial melting of material deep within the Earth, beneath mid-ocean ridges or where a lithospheric plate is subducted. Partial melting also occurs when ascending magmas react with the rocks they pass through en route to the surface. In order to take these ideas further, we need to make a journey. But, before that, we need to say a few words about two other topics of importance for the formation and evolution of the continental crust: granites and their formation, and the change in form associated with *metamorphism*.

About granites and continents

By definition, granites are rocks formed from melts which have cooled relatively slowly at depth in the Earth. For this reason the minerals they contain have formed relatively large crystals. The essential minerals in granites are quartz, potassium feldspar, and plagioclase feldspar with relatively small amounts of biotite or muscovite mica and/or amphibole. Granite rocks form a major part of the continental crust of the Earth, occurring as both relatively small bodies and as the very large plutons now often exposed as the cores of great mountain ranges. Currently known only on planet Earth, granite is the most abundant 'basement' rock of our planet, underlying the relatively thin sedimentary veneer of the continents. So, why is granite so abundant?

The answers to this question have been provided by decades of experiments at high temperatures along with our evolving ideas about the dynamic processes associated with plate tectonics. The experimental work has shown that magmas will evolve so that the last melt compositions (products of differentiation of the kind associated with the Bowen Reaction Series) will be granite melts. Conversely, the first magma compositions to be produced on partial melting of many rock types will be granitic. Such magmas have either evolved or been formed by melting at or near what is termed a *eutectic point* (eutectic being a word derived from the Greek meaning 'easily melting'). This is a mixture of minerals in fixed proportions that melts (or solidifies) at a single temperature which is a lower melting temperature than that of the separate minerals, or any other mixture of them.

In the context of plate tectonics, the magmas formed by partial melting of subducted lithosphere or of material at depth in the continental crust will be granitic in most cases (see Figure 10). A very important finding from experiments is that the presence of small amounts of water can greatly lower the temperature at which melting can occur (by several hundred degrees) such that melting can occur in the deeper parts of the continental crust. The melts formed at depth will be less dense than the surrounding rocks and slowly rise towards the Earth's surface to produce volcanoes if they do reach the surface, or if not, to solidify as plutons. Granites are classified on the basis of their detailed mineralogy, which in turn provides clues as to the processes by which they have formed. For example, whether a granite has been formed from melting of an igneous or a sedimentary source rock (hence *I-type* and *S-type* granites).

Metamorphism

When rocks are subjected to increases in temperature or pressure and temperature, changes occur in the minerals they contain. This may involve changes in the sizes and shapes of mineral grains without the minerals themselves changing. A good example here is that of a marble produced by the metamorphism of a pure limestone. In both rocks, the only mineral present is calcite ($CaCO_3$) but the calcite grains may be coarser and more closely interlocking and compact in the marble than in the limestone. Marble has been a favoured raw material for architects and sculptors since ancient times because it is softer and more homogeneous than many alternatives. The material favoured by Michelangelo for his finest sculptures was a pure white marble from Carrara in the Italian Alps. Most metamorphism, however, involves changes in the minerals present in the rock. There are a number of kinds of metamorphism; two of the most important are contact metamorphism and regional metamorphism. As the name suggests, contact metamorphism occurs where a magma is brought into contact with the rocks into which it has been emplaced ('intruded'), thus raising temperatures and, in some cases, introducing new chemical components to form new minerals. These changes are only local and diminish away from the 'intrusion'.

Regional metamorphism has affected very large areas of the crust, notably where the rocks are parts of either present-day or ancient mountain belts. These are rocks that have been subjected to increases in both temperature (typically somewhere between 100°C and roughly 900°C, above which rocks begin to melt) and pressure. It was in the ancient mountain belts of the Scottish highlands that the British geologist George Barrow established, in work published in 1893, the idea of metamorphic zones (later known as 'Barrovian' zones). Here, each zone is an area of rocks characterized by the appearance of a key mineral or minerals produced by changes in the original rock caused by the higher temperatures and pressures of metamorphism. With increasing temperatures and pressures, which were later attributed to increasing depth of burial, Barrow identified zones he named chlorite, biotite, garnet, staurolite, kyanite, and sillimanite after the key minerals in rocks that were originally mudstones. Kyanite and sillimanite are typical metamorphic minerals. They are minerals of the same chemical composition (Al_2SiO_5) but different crystal structures. Whereas sillimanite only forms at relatively high temperatures, kyanite forms at relatively high pressures.

Metamorphic zones were later defined in much more detail by the great Finnish geologist Pentti Eskola who identified the ranges of pressure and temperature leading to different groups ('assemblages') of minerals being produced from the same initial rock composition. He observed the sequence of minerals produced during metamorphism of basaltic rocks, leading to the concept of metamorphic *facies*. At a specific range of temperature and pressure (a specific facies) a metamorphosed basaltic rock will contain a specific assemblage of minerals. If the temperature and/or pressure changes, the mineral assemblage changes. Most regional metamorphism is also accompanied by deformation due to tectonic stresses, which produce an alignment of the minerals formed as a result of reactions due to increased temperature and pressure. This can produce rocks such as slates, where the alignment of minerals means that the rock can be easily broken ('cleaved') in one direction to produce the very thin flat sheets widely used as roofing materials. Slates are produced by the regional metamorphism of mudstones.

The large-scale processes involved in regional metamorphism are now understood in the framework of plate tectonics and its role in mountain building. In what are amongst the most remarkable achievements in geological research over the past half-century, it has been possible to unravel the detailed geological histories of great mountain belts such as the Himalayas. This has been possible by combining large-scale field studies with the information gained from laboratory investigations of the stabilities of metamorphic minerals and their uses as indicators of temperatures and pressures at their time of formation (as so-called 'geothermometers' and 'geobarometers'). Another important contribution to the 'detective work' needed to unravel these histories has come from advances in

Minerals

the age dating of minerals. A good example of such work involves the mineral zircon ($ZrSiO_4$) which is a common minor mineral in many igneous and metamorphic rocks. This mineral contains trace amounts of uranium and thorium, which are slowly decaying radioactive elements used in radiometric dating (i.e. determining the age of the mineral since formation by determining the extent of such radioactive decay).

Journey to the centre of the Earth

Although Hollywood movies and the classic fiction of Jules Verne might suggest that humans can directly explore the Earth's interior, the deepest mines have barely reached 4 km beneath the surface and deepest boreholes only 11 km, whereas the average distance from the surface to the centre of the Earth is 6,371 km. Increases in temperature and pressure with depth impose practical limitations, with pressures at the centre of the Earth calculated to be nearly 4 million times greater than pressure at the surface and temperatures estimated at 4,300°C. Therefore, our understanding of the interior is based on indirect evidence, and on experiments or computer modelling.

The most important evidence comes from the study of earthquake waves. The velocity at which an earthquake wave travels through a material such as a mineral is a fundamental property which can be measured in the laboratory; it is a property directly related to its 'stiffness' and its density. The stiffer the material the faster the earthquake (seismic) wave velocity. The waves from a major earthquake pass right through the Earth and are detected at hundreds of seismic stations scattered over the globe. The data acquired from these stations are used to determine how seismic velocities change with depth in the Earth. As we would expect, there is an increase in seismic velocity with depth attributable to the increase in density and stiffness as the materials are compacted under the enormous pressures which must prevail at greater depths.

Two other key observations from such studies are that there are relatively abrupt changes in velocity with depth, showing that the Earth is a layered body made up of concentric shells, and that the layer that surrounds a solid inner core is actually a liquid outer core. We know this because a certain type of seismic wave (a shear or 'S' wave) is not transmitted through it. The shells identified in this way and their thicknesses are shown in Figure 11a and, in Figure 11b, how seismic wave velocity changes with depth in the Earth is also shown. The shells consist of a solid inner core and liquid outer core, a solid mantle divided into a lower mantle, a transition zone and an upper mantle, the uppermost part of which is plastic (the asthenosphere), all overlain by a rigid lithosphere (the upper part of which is called the crust). The crust is also

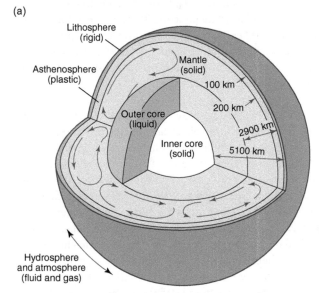

11. The interior of the Earth: (a) showing the concentric shell structure

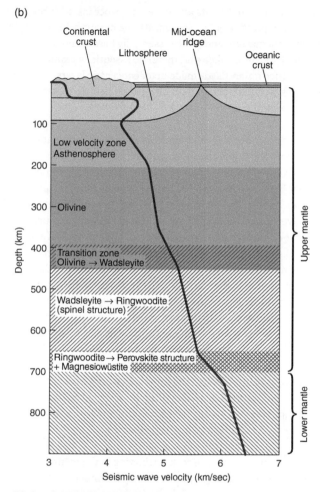

(b)

Continental crust

Lithosphere

Mid-ocean ridge

Oceanic crust

Low velocity zone
Asthenosphere

Olivine

Transition zone
Olivine → Wadsleyite

Wadsleyite → Ringwoodite
(spinel structure)

Ringwoodite → Perovskite structure
+ Magnesiowüstite

Upper mantle

Lower mantle

Depth (km)

Seismic wave velocity (km/sec)

11. (b) showing phase transformations in the mantle and evidence
from changes in seismic wave velocities

different on the continents and beneath the oceans, with the average compositions of rocks beneath the oceans being richer in iron and magnesium and hence essentially basalt, and the continents richer in silica and therefore essentially granite. Given this evidence for an Earth made up of concentric layers, what are the changes in mineralogy that take place when we pass from one layer into another?

The seismic properties offer valuable information to guide our ideas about the minerals making up the rocks that form these successive shells. Other clues come from a number of sources. In what has been described as perhaps the most important experiment of 18th-century science, Henry Cavendish determined the universal gravitational constant. He did this by measuring the force of attraction between objects of large mass (large lead spheres). From this and the Earth's radius, determined from measuring the curvature of the Earth, it was possible to determine the mean density of the Earth as 5.52 (grams per cubic centimetre). As rocks at the surface have an average density of 2.7, much of the interior must have a density greater than 6.0, in line with our ideas of an internally layered body. Although we cannot directly sample rocks from depths greater than a few kilometres (km), samples of rocks from the upper mantle can be carried to the surface by volcanic activity. Commonly these are brought up as lumps of rock termed *xenoliths* (literally 'foreign rocks') in a magma and are dominated by magnesium and iron silicate minerals of the olivine and pyroxene families.

An important example of a rock originating in the mantle and brought to the surface by volcanic activity is kimberlite. This rock is also important because it is the main source of diamonds (and is named after the famous Kimberley diamond mines in South Africa). Coming from depths of around 150 km, these rocks contain a wide range of xenoliths, including olivines and pyroxenes from the mantle, as well as material entrained from surrounding rocks at shallower depths. Kimberlites are mostly found in the form of volcanic chimney structures (called *pipes*) which record the routes along which they blasted a way to the surface, probably powered by admixed volatile gases made up largely of steam and carbon dioxide. The presence of diamonds in some kimberlites further testifies to their origins at great depths, diamond being a dense and compact form of carbon that can only be synthesized at very high pressures, as we have already discussed (see Figure 1c). The origins of kimberlites, and why only some of them contain diamonds, remains something of an enigma. This is partly because there are no examples of such rocks actually being formed at the present day.

More indirect clues to the minerals making up the interior of our planet come from the study of meteorites. These fragments left over from the material that accreted ('clumped together') to form the rocky planets of our Solar System sometimes survive the journey from elsewhere in the solar system, such as the asteroid belt, and crash to Earth. There are two main types of meteorites. The stony meteorites are made up of silicate minerals, mostly olivine and pyroxene, and so resemble the rocks of our upper mantle. The iron meteorites are largely made up of iron alloyed with some nickel. They provide strong indirect evidence for an Earth's core that is dominantly composed of iron, probably with some nickel.

There are two additional complications we need to address in any attempt to understand the make-up of the core; the first is the seismic evidence for a liquid outer core which is also required to explain the Earth's magnetic field. It is suggested that the magnetic field arises from the motions of a layer of molten iron acting as a dynamo. Only in this way can the complexities of the Earth's field be explained. These complexities include changes in the direction and intensity of the magnetic field at Earth's surface from place to place and on a timescale of only hundreds of years, and the reversals of the polarity of the Earth's magnetic field (magnetic north becoming south and vice versa) which have happened on numerous occasions over geological time (*magnetic reversals*). The second complication is that the seismic evidence points to an inner core which has a density somewhat less than iron itself, so that there must be another element (or elements) present in the core. The identity of the light element contribution to the core is still a matter of debate; sulphur, silicon, and potassium have all been suggested.

Experiments at high pressures

One of the most important areas of research on minerals concerns experiments at high pressures. Because of the links with geophysics and with the physical properties of materials, this area is a major part of the field often termed *mineral physics*. The importance of this area will be obvious, given that the overwhelming majority of minerals are at high pressures in the Earth's interior. Topics of particular interest are pressure-induced changes at the boundaries between the different 'shells' making up the regions of the Earth's interior, and changes in mineralogy when lithospheric material is subducted into the mantle at a destructive plate boundary. Less dramatic, but no less important, are the measurements of physical properties such as seismic wave velocities at high pressures in suspected mantle minerals.

There is a distinguished history of research in high pressure mineralogy closely linked with materials research at high pressures. The earlier work involved equipment such as a hydraulic press forcing anvil jaws together, and with the sample surrounded by a furnace to simulate the high temperatures at depth in the Earth (see Figure 12a). Such equipment and more sophisticated rigs developed from the early apparatus were capable of simulating conditions at depths of about 300 km, roughly halfway down through the upper mantle. Much greater pressures, reaching those of the core, were also achieved for very brief periods of time by using explosives to generate shock waves, or by firing a bullet into a mineral target. Although the compression in such experiments lasts for only a few millionths of a second, it is enough time to determine density, pressure, and the parameters that provide information on seismic wave velocities.

Recent decades have seen great advances in high pressure experimentation. These have been particularly associated with a device called the diamond anvil cell (see Figure 12b). Small enough to be held in one hand, this device contains two gem quality diamonds, cut so that two perfectly flat faces can be mounted face to face with a microscopic mineral sample between them. A little thumbscrew is then tightened to apply the pressure. Because all of the pressure is applied to a tiny area of sample, enormous pressures

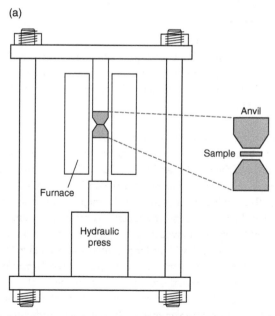

(a)

Anvil

Sample

Furnace

Hydraulic
press

12. High pressure experiments: (a) a simplified drawing of a 'traditional' high pressure rig

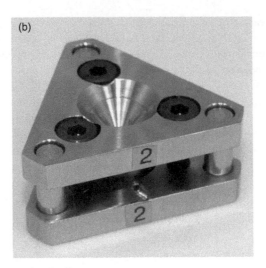

12. (b) a diamond cell

(more than 3 million times atmospheric pressure) can be achieved. Visible light, X-rays, or other forms of radiation can be transmitted through the diamonds and through the sample, which can also be heated using a laser beam.

High pressure experiments, such as those involving the diamond cell, have been especially successful in providing ideas as to the mineralogy of the mantle, and the causes of the discontinuous changes in seismic wave velocities with depth (as seen in Figure 11b). Olivine is believed to be the major component of the upper mantle; some estimates suggest it comprises as much as 58 per cent of the mineral inventory. A sample of olivine compressed in a diamond anvil cell will undergo changes associated with a reorganization of its constituent atoms (*phase transformations*). In such transformations, there is no change in chemical composition, but there is a rearrangement of the atoms to produce a denser and more compact crystal structure. At a pressure equivalent to a depth in the Earth of 410 km, such a transition takes place to a mineral phase called wadsleyite, β-$(Mg,Fe)_2SiO_4$, or simply the 'β phase'.

This transformation to a more compact structure involves a 6 per cent increase in density. Experiments show that wadsleyite then transforms at a pressure equivalent to 500–550 km depth to a mineral called ringwoodite, also referred to as γ-$(Mg,Fe)_2SiO_4$. This phase has the same structure as the mineral spinel, $MgAl_2O_4$, but there is only a small density increase of ~2 per cent associated with this transition, an increase which is not generally enough to produce a resolvable seismic discontinuity (see Figure 11b). Wadsleyite and ringwoodite are found in meteorites, adding further weight to the idea that they might be important phases in the mantle (and providing the natural occurrences essential for them to be given mineral names).

At 660 km depth in the Earth, there is another major discontinuity with a large increase in both seismic wave velocity and in density (5 per cent). Again, experiments at pressures equivalent to this depth show a change in mineralogy, but this time ringwoodite (spinel) changes to a mixture of two phases. One is a silicate, $(Mg,Fe)SiO_3$, with the same structure as the mineral perovskite, $CaTiO_3$, and the other is not a silicate but a simple iron and magnesium oxide mineral called magnesiowüstite, $(Mg,Fe)O$, related to periclase (MgO) and having the very simple and compact 'halite structure' found also in galena (see Figure 1a). A point to emphasize here is that these are experimentally verified phase transformations. They occur at pressures equivalent to exactly where there are abrupt changes in seismic wave velocities within the Earth. Although it cannot be proved beyond all doubt, this strongly suggests that the seismic discontinuities at 410 km and 660 km depth are caused by changes in crystal structure rather than a reaction leading to an entirely new assemblage of mantle minerals.

The high pressure phase transformations of olivine, commonly referred to as the olivine → 'spinel' transition (despite the role of wadsleyite, which actually has a different structure to spinel) are amongst the most important in all Earth sciences. Olivine, $(Mg,Fe)SiO_4$, an 'island silicate' mineral, has quite an open structure in which the SiO_4 tetrahedral units do not share any of their oxygens with other tetrahedra but are held together by Mg and Fe atoms. The spinel structure of γ-$(Mg,Fe)_2SiO_4$ is more dense and compact; it takes up about 8 per cent less space than olivine. The transformations to a mixture of silicate perovskite and magnesiowüstite, represent further compaction. This involves certain of their atoms being bonded to, and therefore surrounded by, a larger number of oxygen atoms (which are more compressible).

Although the transformations of olivine are the best known and most researched of those believed to involve mantle minerals, they are certainly not the only such possible transformations. Garnets and pyroxenes are believed to be important phases in the upper mantle (possibly comprising 10–20 per cent) and these transform to denser phases; for example, at pressures equivalent to depths of 350–450 km, pyroxenes transform to a garnet-structured mineral known as majorite garnet with a density increase of about 6 per cent. This majorite garnet transforms to a perovskite structure at similar pressures to the transformation of ringwoodite (spinel) to perovskite (in this case equivalent to a depth of 650–680 km). Each of these transformations entails a progressive increase in the number of oxygen atoms surrounding the silicon atoms from 4 to 6, and hence a more compact, denser structure.

There is one other important part of the story of high pressure phase transformations. It concerns the mechanism and the kinetics (i.e. the 'speed') at which a crucial phase transformation such as that of olivine → 'spinel' takes place. At destructive plate margins, olivine-bearing rocks are dragged down into the mantle (subducted) and, at some depth, the olivine will transform to a 'spinel'-type high pressure form of greater density.

Ultrafine grained material forming as a transient product of this reaction (before grain growth occurs) may facilitate deformation by a process of sliding along grain boundaries, producing fault zones. This process is thought to be responsible for deep-focus earthquakes. The point at which a high pressure phase transformation starts is determined by both the depth (pressure) and temperature. It is regarded as unlikely that nucleation and growth of the new phase would occur at temperatures lower than about 700°C, the expected temperature at the top of the transition zone. A subducted slab of lithosphere is likely to be colder than the surrounding mantle rocks, and it is likely that a metastable wedge of olivine-containing rock may persist well into the transition zone. As the slab heats up, the transition may occur very rapidly, with the weakening due to the formation of ultrafine reaction products leading to a release of energy as seismic waves, detected as deep focus earthquakes in the downgoing slab of lithosphere.

The deeper parts of the Earth's interior have an important role in the plate tectonic cycle. Although the mantle is solid rock, it is hot and weak enough below the lithosphere to flow like a viscous liquid. Heat from the decay of radioactive elements and from the still molten outer core can cause local heating at depth. The heated mass of rock expands, becomes less dense, and rises very slowly. To compensate for this rising mass, rock that is cooler and denser must sink downwards in a process of convection (see the arrows in Figure 10). The very large scale of these processes is indicated by seismic data used to 'map' the mantle. In seismic tomography, data can be obtained for the Earth's deep interior rather as we now obtain data on human organs using a body scanner. In fact, the mantle is found to be very heterogeneous. Also, the measured rate at which heat reaches the Earth's surface can only be accounted for if this heat comes from the deep mantle and core.

There is still much to learn about the deeper parts of the Earth and the fate of subducted slabs of lithosphere. For example, just above the core is a layer, perhaps 100–400 km thick, termed the D″ layer. It is very heterogeneous, both vertically and laterally, and suggested to be the remains of lithospheric slabs that have sunk to the base of the mantle. Here, mixing between molten iron from the core and high-pressure silicates could take place, and experiments suggest vigorous reactions producing a mixture of magnesium perovskite, wüstite (FeO), plus a high pressure form of silica, and iron silicide (FeSi).

Convection in the mantle is regarded as the main driving force for the creation and destruction of lithospheric plates, causing the upwelling of magma to form new crust at spreading centres, and its destruction at subduction zones. Even a long way from these plate boundaries, there appear to be so-called mantle plumes, where hot material slowly rises from depth and causes a 'hot spot' within a plate rather than at a plate margin. The Hawaiian Islands, located in the middle of the Pacific Plate, are an example of hot spot volcanism.

The D″ layer is also thought to be the source of the mantle plumes that give rise to hot spots, the deep mantle source of such plumes contrasting, it has been proposed, with magmas at ocean ridges fed from the uppermost mantle.

The role of minerals in the plate tectonic cycle is evident from the above discussion. The unending formation and the destruction of the lithosphere, driven by energy from the interior of the Earth, is part of a cycling of rocks and minerals that is characteristic of our dynamic planet. But this is not the whole story. To complete the story of the cycling of minerals, we now need to consider transformations driven largely by energy coming from the sun.

Chapter 4
Earth's surface and the cycling of minerals

Although the movement of tectonic plates seems to us very slow, with 'continents' drifting at most a few centimetres in a year, these processes are responsible for most of the drama we associate with the 'dynamic Earth'. This includes earthquakes, tsunamis, and volcanic eruptions. As we have seen, there is a cycling of minerals taking place on a very large scale as part of the plate tectonic cycle, with material brought up from the mantle forming new ocean floor and that material, in turn, eventually being consumed by subduction. There is also a cycling of minerals that is less dramatic but of great importance to our understanding of Planet Earth. It is also part of our common human experience, wherever we live on Earth, and has great practical as well as theoretical importance. This is the cycling of minerals at or very near the surface of the Earth, cycling particularly associated with the weathering and erosion by rain, wind, and frost action of rocks exposed at the surface. Some of the minerals in those rocks may be dissolved during weathering, others may be transported in the flowing water of streams and rivers, by glaciers, or as fine mineral dusts in the atmosphere, eventually being deposited elsewhere as sediments. This is part of what geologists call the *rock cycle* (which is also necessarily a 'mineral cycle'), but one involving the rocks of the Earth's crust.

As seen in Figure 13, we can think of the complete rock cycle as beginning with igneous rocks formed by crystallization from melts; when exposed at the Earth's surface, these rocks will weather and be eroded and transported to form sediments. These unconsolidated sediments can then be buried and slowly transformed into solid sedimentary rocks (the process of *lithification*). Deep burial of such rocks can subject them to greatly increased pressures and temperatures, causing their mineral components to change in various ways in the processes of *metamorphism* already discussed in Chapter 3. As noted there, this may involve the minerals in these rocks changing in the sizes and shapes of their crystals, or more importantly, transforming into new minerals. Eventually, they may be so deeply buried as to start melting, taking the cycle full circle back to igneous rocks. As shown in Figure 13, it is also possible for the cycle to be broken with a return to an earlier point in the cycle; so sedimentary rocks can be exposed to weathering and erosion at the Earth's surface to form new sediments, as can metamorphic rocks.

The overall rock cycle can be thought of as having two parts. The first is driven by the heat coming from the Earth's interior and involves interactions between the mantle and crust. The minerals formed in this part of the cycle crystallize from melts (*magmas*) or very hot waters (*hydrothermal waters*). The second is driven primarily by the heat from the Sun and involves interactions between exposed crust and the waters of the *hydrosphere* or gases of the *atmosphere*. As we note above, and in Figure 13, the rocks formed in this part of the cycle are deposited at Earth's surface temperatures as sediments. These may be detrital in origin (such as the grains of quartz in a sandstone), may come from evaporation of waters containing dissolved mineral components (such as the rock-salt in a salt bed), or have a biological origin (such as the calcite originally deposited by the coral animals in a 'reef' limestone).

The biological processes involved in the formation of the calcium carbonate minerals (calcite and aragonite) making up many

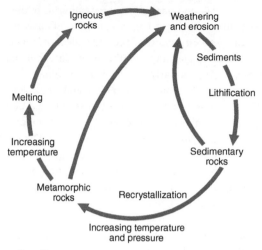

13. The rock cycle

limestones are a reminder of the importance of the *biosphere* in the cycling of minerals at the Earth's surface. It is also true that minerals have a vital role in sustaining life on Earth. For this reason, the region of the surface and near-surface that is essential to sustaining life is now called the *critical zone*. More will be said about minerals and the biosphere in Chapter 5.

Weathering

The weathering of minerals and rocks at the surface of the Earth depends on a number of factors. As well as the nature of the mineral itself, these include climate, vegetation, micro- and macro-organisms such as bacteria or earthworms, slope of the ground, and time since exposure of the mineral at the Earth's surface.

A good example of mineral weathering is provided by the products of weathering of a typical granite, probably the best known of all igneous rocks. The minerals which comprise a typical granite are feldspars, micas, and quartz. During weathering of the orthoclase feldspar ($KAlSi_3O_8$), minor amounts of potassium (K) and silica (SiO_2) go into solution and the feldspar reacts to form a potassium-bearing clay mineral; the plagioclase feldspar (($Na,Ca)Al_{1-2}Si_{3-2}O_8$) similarly may form a clay mineral but one containing sodium (Na) and calcium (Ca), and silica along with Na and Ca go into solution; the micas (muscovite and biotite) form clay minerals plus haematite (Fe_2O_3), soluble magnesium (Mg) and silica, and minor soluble potassium (K). The quartz (SiO_2) will survive even intense weathering largely unaltered. The golden sands of beaches in granite areas (such as western Cornwall, England) are the products of this survival, being almost entirely quartz grains.

The example of granite weathering highlights the importance of clay minerals as breakdown products of feldspars and micas. They can also form from the breakdown of other silicates such as the olivines, pyroxenes, and amphiboles. Clay minerals are silicates containing aluminium and also bonded water in their structures, and, in most cases, other elements such as sodium, potassium, calcium, magnesium, or iron. All clay minerals are layer or '*phyllo*' silicates (see Table 1; and Figure 14) with crystal structures in which SiO_4 tetrahedral units share oxygen atoms to form layers, and AlO_6 or MgO_6 octahedral units also link together to form layers. Various combinations of layer sequences are possible with, in some cases, charged atoms (*ions*) such as K^+, Ca^{2+}, Mg^{2+} between certain of the layers (as *interstitial ions*) to bond them together. Clay minerals are always of very small particle size and their structures and compositions give them distinctive properties. As well as being important components of soils, these properties make clays the critical raw materials for a range of industries, the oldest of which is the manufacture of pottery. As early in human history as 9000 BC, clays were being fired to make pottery. Brickmaking and other ceramic arts followed by 3500 BC. Modern industries use clay minerals in numerous products. For example, kaolinite ($Al_4Si_4O_{10}(OH)_8$), also known as china clay, is not only used in manufacturing fine porcelain but in providing the coating on the surface of the high quality paper used in glossy magazines. Another important property of many clay minerals is a capacity to take up impurities such as ions or molecules from solution. This gives clay minerals in soils a key role as the carriers of plant nutrients, and also applications in certain industrial processes where impurities need to be removed from contaminated liquids. One of the oldest such processes is that of *fulling*, which is the removal of grease from wool or other organic fibres. Because of this application, natural deposits of the clays used in this way are known as 'fuller's earth'.

The breakdown of silicate minerals dominates the weathering stage of the rock cycle. The only other minerals that compete with silicates in terms of crustal abundance are the carbonates of limestone rocks. Limestones are made up almost entirely of calcium carbonate (calcite). Magnesium may replace half of the calcium in calcite to give dolomite, $CaMg(CO_3)_2$, the name given to the mineral, a rock made of that mineral (more correctly called dolostone) and, incidentally, the 'Dolomites' mountain range in the Italian Alps built mainly of that rock. The behaviour of carbonates during weathering is controlled by their solubility in rainwater. Uptake by rainwater of carbon dioxide from the atmosphere means that it is not pure H_2O but weak carbonic acid (H_2CO_3) which can dissolve carbonates. This is the origin of the many distinctive features of limestone landscapes which include spectacular cave systems formed by limestone dissolution. Although much less abundant, one other group of minerals must be mentioned in any discussion of weathering; these are the sulphides, especially the iron sulphide, pyrite (FeS_2).

Sulphide minerals occur in substantial amounts in the wastes from the mining of many metals and the mining of coal. They can also be minor components in common rocks, especially certain shales. Sulphides are not stable when exposed to oxygen or oxygenated waters at Earth's surface; they break down to produce sulphuric

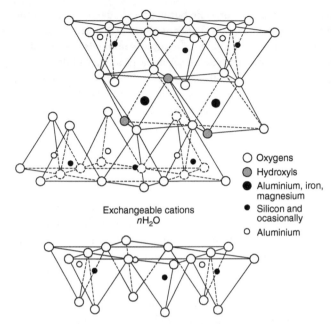

O Oxygens
◐ Hydroxyls
● Aluminium, iron, magnesium
• Silicon and ocasionally
○ Aluminium

Exchangeable cations
nH$_2$O

14. The crystal structure of the clay mineral montmorillonite (Al$_4$(Si$_4$O$_{10}$)$_2$(OH)$_4$.nH$_2$O); a fragment showing the relationships between layers made up of tetrahedrally and octahedrally coordinated atoms and the interlayer which normally contains weakly held water ('nH$_2$O') which can be exchanged for other molecules

acid and, initially, to take the iron or other metals into solution in the acidified waters. The name given to this phenomenon when the sulphide mineral source is minewastes is *Acid Mine Drainage* (AMD); when the source is the exposure of certain common rocks, it is *Acid Rock Drainage* (ARD). Nearer to the source, the waters produced can be highly acidic, comparable to automobile battery acid, or even more extreme. Bodies of water filling former open pit mine workings can be sufficiently acid to dissolve discarded metal objects. When the waters of AMD systems mix with other surface waters, they cause acidification, killing off aquatic life and blighting whole stretches of waterways.

It has been estimated that over 10,000 miles (~16,000 km) of streams and rivers in the USA alone have been seriously polluted by AMD, a legacy from the mining of metals and coal. As well as acidification, such pollution involves large amounts of very fine particulate mineral matter. This is precipitated as the waters downstream become less acid by mixing with unpolluted rivers and streams and can no longer retain metals in solution. For example, the iron held in solution in highly acidic waters will precipitate as iron hydroxide minerals such as orange-red goethite ($FeOOH$). It is the fine precipitates of such minerals (including examples of the nanominerals discussed below) that cause the waters in many polluted areas to turn into 'red rivers'. These particles of iron minerals can also transport downstream other elements trapped within their structures or attached to their surfaces, including highly toxic elements such as arsenic or uranium. The uptake of other elements in this way is further considered when we discuss the nature of mineral surfaces later in this chapter.

Nanominerals and nanoparticles

Much of the material derived from the breakdown of the common minerals is of very fine particle size. In some cases, the particle size can be measured in *nanometres* (nm) where one nanometre is one thousand millionth of a metre. This includes much clay mineral matter and also may include iron oxides or hydroxides such as goethite ($FeOOH$) or the aluminium hydroxides ($AlOOH$). The study of these materials, now called 'nanominerals', is a relatively recent development in mineralogy. This is partly because their geological importance was not fully appreciated, and partly because techniques to study them were not available. The experiments now possible at synchrotron laboratories, along with scanning probe microscopy methods and advances in electron microscopy, have provided powerful new investigative methods. Two other important aspects of nanominerals need to be discussed here. The first is that much evidence points to nanomaterials (and nanominerals) exhibiting properties that are distinctly different from those of their larger particle size equivalents. The second is that manufactured nanomaterials are now very widely used in industry, prompting much research into their properties, as well as concerns about the impact these materials may have when released into the environment.

Michael Hochella of Virginia Tech (USA), a pioneer of nanomineral studies, has proposed a nanoscale mineral classification scheme. All nanoscale minerals must have at least one dimension in the nano-range (between 1 and 100 nm) but may occur as *nanosheets* or *nanorods* as well as nanoparticles. A very important further distinction is drawn between *mineral nanoparticles* and *nanominerals*: the former can also exist in sizes that exceed the nano-range (possibly up to the largest dimensions found in minerals) whereas the latter exist only as one of the three types of nanoscale minerals noted above. A very good example of a nanomineral is ferrihydrite ($Fe_{4-5}(OH,O)_{12}$), a mineral that is very common in soils and natural waters. It has never been found as particles larger than 20 nm, and is typically 10 nm or less in diameter. It is very probable that many such nanominerals remain to be discovered.

The properties of nanoparticles can be dramatically different compared to their larger-size equivalents. For example, solubility in water may be orders of magnitude greater for nanoparticles compared with larger particles of the same mineral. It may also be that the crystal structure of the mineral that is most stable at atmospheric temperature and pressure in the nanoparticle size range is not the most stable at a larger size. This is the case for the titanium dioxide (TiO_2) minerals where there are three different crystal structure forms (*polymorphs*) and hence minerals: rutile, anatase, and brookite. Here, anatase, rather than rutile, is the stable phase only for nanoparticle sized grains. Even where particle size does not influence crystal structure, it may influence the preferred crystal shape (e.g. cube versus octahedron or tetrahedron). So, size really does matter!

In part, at least, these differences arise because so many of the atoms in a nanoparticle are at or very near the surface (a 10 nm diameter cube crystal will have 16 per cent of its atoms at the surface). One consequence of this is the very large surface area presented to the environment (the air or natural waters) for reaction and interaction. In particular, processes discussed in detail below where contaminants in waters are taken up (*sorbed*) by the surfaces of nanoparticles are very important. Toxic elements such as arsenic, lead, cadmium, or radioactive contaminants such as uranium may become attached to these surfaces and transported considerable distances in flowing water. Nanominerals or nanoparticles may also coat larger grains such as those of quartz, causing their surfaces to be active for sorption, or alternatively nanoscale films may coat much larger grains of minerals such as feldspars, inhibiting their weathering.

The subject of 'nanomineralogy' is so new that we have no real idea of the total quantities of nanoparticles in the Earth's surface and near-surface environments, or how they are distributed. It is suspected that most are to be found in the oceans. In the oceans there is also an important role played by nanoparticles of iron minerals. Phytoplankton, important contributors to ocean ecosystems, require iron for their metabolism; this iron had always been assumed to come via input from rivers, particularly the great rivers such as the Amazon. Now it appears that airborne dusts carrying iron minerals provide an input that far exceeds that from rivers.

More surprising occurrences of nanoparticles are those found in minute concentrations in stony meteorites and interplanetary dusts; the latter includes nanodiamonds. They average 3 nm in diameter, although grains as small as 1nm and, therefore, containing fewer than 150 carbon atoms have been observed. Nanodiamonds are believed to represent pre-solar dust, possibly forming in supernovae, although it has also been suggested that they could form directly in the solar nebula. Our unmanned exploration of nearby planets has also revealed evidence of nanoscale ferric oxides on the surface of Mars. Also surprising, and a link to our earlier discussions of 'Deep Earth' (in Chapter 3) are the proposed examples of nanoscale particles of the very high pressure minerals wadsleyite and ringwoodite. Here a role is envisaged for nanominerals in the generation of deep focus earthquakes, those at 300–700 km depth in the mantle. The mineral nanoparticles are envisaged as filling 'anticracks' which are planes of weakness that do not need to dilate to create empty space, unlike a normal crack. These nanoparticles can easily move past each other without mechanical shearing of individual grains, so that high pressure does not restrict such movement. There is also evidence that the mechanical properties of nanoparticles can play an important role at shallow depths, influencing fault mechanics, and that the mineral particle size in the nano-range may affect compressibility.

The upsurge of interest in natural nanoparticles and nanominerals is partly a result of the development of synthetic nanomaterials and of nanotechnology. Beginning particularly with the nanoscale forms of carbon, such as the spherical cluster of 60 carbon atoms (the C_{60} phase named *buckminsterfullerene* or the 'buckyball' for short; see Figure 1d) and carbon nanotubes, numerous elements and compounds are now synthesized as 'nano' materials for use in industry. Applications range from electronics to environmental clean-up, and medicines to cutting tools. The global nanotechnology market is already estimated to be worth several trillion dollars annually. However, although many will end up in landfills or similar disposal facilities, the impact of these synthetic nanomaterials on the environment is very poorly understood. Indeed, the extent to which both natural and synthetic nanomaterials can enter into biological systems (their so-called 'bioavailability') is still poorly understood. This is a subject we return to in Chapter 5.

Something in the air

Probably the most important scientific issue of the 21st century is global warming and its potential impact on sea levels and climate worldwide. The greenhouse gases, particularly carbon dioxide (CO_2), have increased considerably in concentration in the Earth's atmosphere since the beginning of the industrial revolution as a result of the burning of fossil fuels, especially coal, gas, and oil. The greenhouse gases absorb infrared radiation and, therefore, warm air near to the Earth's surface, so as to cause an increase in overall average temperatures. Potentially catastrophic consequences of even a modest increase in average temperatures have led governments to introduce policies restricting greenhouse gas emissions. Possible consequences of global warming include melting of polar icecaps, with rises in sea level which could flood many coastal cities, towns, and rural areas, and more frequent extreme weather phenomena such as heatwaves and droughts, or storms and hurricanes.

You might think that minerals have nothing whatever to do with global climate, but that is not the case. Every year, 1,500–2,600 million metric tons of 'dust' is transported around the globe. Much of this dust (an important component of the transported solid particles and liquid droplets more generally called *aerosols*) is mineral matter picked up by winds from soils or sands which are themselves the erosional products of rock weathering. Some is 'sooty' matter from burning of fossil fuels or of agricultural waste (*biomass*) burned as part of a crop cycle. Studies of aerosol particles using the electron microscope (TEM) show that, not surprisingly, the minerals present are dominantly the common rock-forming minerals such as quartz, feldspars, micas, clays, iron oxides, and hydroxides. As well as the sooty materials which may be present as coatings on mineral grains, common components of aerosols include various salts, commonly crystallized from solutions such as ocean spray and, of course, ice as well as water droplets.

Aerosols are of interest to us for a number of reasons. In what are fortunately rare cases, airborne mineral matter can pose a significant health hazard. An example of this is in northern and western China, to the west of Beijing. Here deposits of wind-blown silt known as *loess* (with particles of 2–63 microns in diameter) cover an area of over 600,000 square kilometres. Chinese loess has a simple mineralogy: angular, blade-shaped quartz grains make up 60–65 per cent of the silt, with not more than 12–15 per cent clay minerals plus minor feldspars, micas, and carbonates. Dust storms are frequent events in this region and, although detailed epidemiological data are very limited, these dusts are certainly responsible for a much greater incidence of respiratory diseases such as silicosis. Furthermore, particulates mobilized and transported by wind may be carried as far away as north-west India, and cause health problems evidenced by high levels of silicosis which has been linked to mineral dusts by finding silica (quartz) in analysed lung tissues.

Globally, mineral matter in aerosols has a relatively poorly understood influence on climate. Mineral particles can themselves either reflect or absorb radiation from the Sun, or may act as nuclei for the formation of water droplets and, in turn, clouds which may reflect or absorb the Sun's rays. The behaviour of different minerals in these systems will ultimately depend on their fundamental optical properties whereas, of course, mineral grains coated with soot will strongly absorb energy from the Sun. The role of mineral-containing aerosols actually introduces significant uncertainty into climate models.

Surfaces and interfaces

'The boundary is the best place for acquiring knowledge' are not the words of a scientist but of the theologian Paul Tillich, but they are very relevant to the study of minerals. This is because the processes controlling the cycling of minerals at the Earth's surface nearly all occur at the boundary ('interface') between the mineral and its environment, whether that be the gases of the atmosphere, or the waters in rivers, lakes, and oceans, or water trapped in soils or underground. Understanding of interfacial phenomena requires an understanding of the mineral surface at the level of individual atoms.

Numerous methods have been developed for characterizing the surfaces of solids; these include the imaging methods discussed in Chapter 2, and various diffraction methods enabling the relative positions of atoms at the mineral surface to be determined. As with many kinds of experiments, the availability of synchrotron radiation has enabled significant advances. A first question we need to consider in studies of mineral surfaces is whether the arrangement of atoms at the surface is like a simple truncation of the bulk (i.e. as though we were to slice, knifelike, through the mineral along a certain plane of atoms). In fact, surfaces are rarely so simple. There are changes in the positions of the atoms at or near the surface which may involve a substantial reorganization (*reconstruction*) or just a smaller adjustment (*relaxation*). These changes are about forming the most stable arrangement of atoms at the mineral surface, i.e. about minimizing surface energy. Understanding the nature of the clean surface is an important step towards understanding how the mineral may react when exposed to the air, pure water, or water that contains impurities such as metal ions or organic molecules.

The most important interface in the cycling of minerals at the Earth's surface is the mineral–water interface. Here, a mineral may dissolve, releasing its constituent elements into solution. This dissolution may continue until the mineral is consumed, or alternatively, lead to an altered surface layer which may protect the surface from further reaction. Other kinds of reactions may occur at the mineral–water interface, especially reactions involving the uptake of atoms or molecules present as contaminants in the water. These are generally referred to as 'sorption' and can be studied by adding set amounts of the mineral of interest to a solution containing a known amount of the impurity. Separation of the solution from the solid mineral and chemical analysis of the solution provides a measure of how much of the impurity is removed from solution through interaction with the mineral. In this way, uptake can be measured as a function of variables such as initial concentration of the impurity in solution, or how acid (or alkaline) is the water.

Such overall measurements of uptake tell us nothing about what is happening at the mineral surface at the atomic level. That information can be obtained through spectroscopic methods, notably the X-ray absorption spectroscopies such as EXAFS as described in Chapter 2. Good examples of the possible interactions at the mineral surface are those involving metal atoms (present in the solution as 'ions' such as Cu^+, Cd^{2+}, Pb^{2+}) and minerals such as the iron oxides and hydroxides (hematite, Fe_2O_3; goethite, $FeOOH$). The possible types of interactions are illustrated in Figure 15, which shows that, in solution, the metal ions (M) will be surrounded by attached water molecules. 'Sorption' of the metal to the mineral surface can involve retaining those water molecules in what is called an *outer sphere surface complex* or losing some of the water molecules to form an *inner sphere surface complex*. Inner sphere complexation means that the sorbed metal forms a direct chemical bond with an atom at the mineral surface and so is much more tightly held than an outer sphere complex. With terminology suggestive of the dentists surgery, inner sphere complexation with the metal bonding to just one mineral surface atom is called *monodentate*, with two surface atoms *bidentate*, and so on. An important feature of both types of surface complexation is that uptake from solution is limited by sites available on the mineral surface. Once all available surface sites are occupied, no further uptake is possible. That does not apply to the other possibilities for interaction illustrated in Figure 15. They lead to *precipitation* of material from solution, or of material derived from both the solution and the mineral (*co-precipitation*) or to *replacement* of

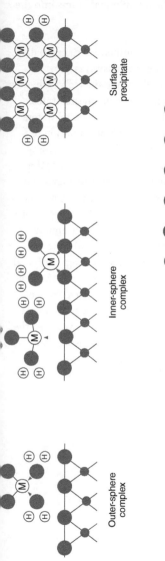

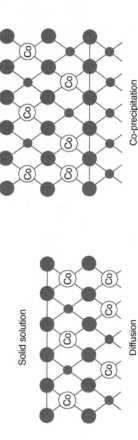

15. **Surface complexation and related interactions with simplified representations of a mineral surface in contact with a liquid containing a metal (M) in solution. The metal is bonded to the oxygen atoms (large grey spheres) of water molecules (hydrogen atoms shown as 'H') as a 'hydration sphere'. Different types of surface complexes, precipitates, or replacement reactions are shown**

atoms in the mineral involving *diffusion* of metal atoms into the surface of the mineral.

You might well ask 'Does all this work on mineral surfaces really matter?' On both global and local scales the answer is 'yes'. Much of the material being transported around the globe, whether in the air or in moving water, is attached to mineral substrates. Reactions between mineral substrates and contaminated waters also lead to the formation of minerals through precipitation or replacement reactions. Locally, these mineral surface processes provide key mechanisms for the transport and dispersal, or the localized concentration of major, minor, and trace elements and molecules including those of organic compounds. These substances include toxic elements such as arsenic, cadmium, lead, or mercury, elements from hazardous radioactive wastes such as uranium, radium, neptunium, technetium, and plutonium, and toxic organic compounds from industrial sources. Modelling of mineral surface processes in this context is essential for assessing the risks of pollution, and for remediation of contaminated areas. In fact, 'sorption' by mineral substrates is one of the methods we can use to remove toxic contaminants (such as arsenic) from water. At the other extreme from a role in protecting living organisms from poisons, mineral surfaces almost certainly played a role in the emergence of life on Earth, an idea we explore in the final chapter of this book.

Chapter 5
Minerals and the living world

When I was a student in the 1960s, apart from the fossil record, the geological and the biological domains were regarded as essentially separate. In our chemistry classes, we were not taught organic chemistry as it was largely regarded as irrelevant for our studies of minerals and rocks. Today the 'geo-bio' interface is one of the most active areas of research in the natural sciences with major research programmes and new journals with titles such as *Geobiology, Geomicrobiology Journal*, and *Global Biogeochemical Cycles*. There are no better examples of 'modern mineralogy' than those provided by the work being done on mineral–microbe interactions, biomineralization, minerals and the human body, and minerals and human health.

Minerals and microbes

The environmental electron microscope picture used in Chapter 2 to illustrate this method of studying minerals (Figure 8) shows several bacteria of the species *Geobacter sulfurreducens* on an iron oxide mineral substrate. These bacteria are not simply sitting on this mineral surface, they interact with it because it is their method of 'respiration', just as much as breathing in oxygen is ours. The transfer of electrons between the microbe and the mineral, in this case, brings about the *reduction* of the iron in this mineral from the ferric (Fe^{3+}) to the ferrous (Fe^{2+}) form. As illustrated in

Figure 16, this can be done by this single-celled organism being directly in contact with the mineral or by an organic molecule (such as a humic compound) 'shuttling' back and forth between the mineral and the microbe, resulting in transfer of the electron. Also involved here is organic matter (in this case it is acetate) which is oxidized to CO_2. This is in effect the 'food' consumed by this organism. A very important consequence of the *microbial reduction* of the iron in this oxide is that it releases the iron into solution (the ferrous iron is soluble unlike the highly insoluble ferric form). This also means that any contaminants trapped within the oxide or attached to its surface will also be released into the surrounding waters.

The reduction of iron by *Geobacter* is just one example of how important mineral–microbe interactions can be for the cycling of the elements in systems at or near the surface of the Earth. Other organisms can bring about the oxidation of ferrous iron in minerals, and there are numerous other examples of microorganisms controlling the reduction–oxidation ('redox') chemistry or other variables in a solution in contact with a mineral, such as the acidity. These changes may be responsible for the formation of minerals or their destruction. Good examples are provided by the iron sulphides. In estuaries or shelf areas just offshore where fine grained sediments have been deposited, a foot or so beneath the surface of the sediment, bacteria (such as the *Thiobacilli*) are active in reducing the dissolved sulphate (SO_4^{2-})—always present in seawater—to sulphide (S^{2-}); this sulphide will then react to form very insoluble iron sulphide minerals. The first formed mineral is mackinawite (FeS) but, with time, the mackinawite reacts further with sulphur to form pyrite (FeS_2). If other metals are present, such as copper or zinc, they will also react to form sulphides.

The reverse of this process can also be promoted by bacteria. In this case, metal sulphide minerals such as pyrite exposed at or near the Earth's surface by uplift and weathering of a sulphide ore deposit may break down to form sulphates (such as $FeSO_4.7H_2O$,

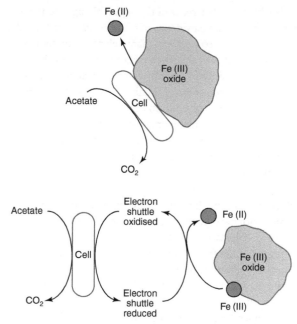

16. **A schematic diagram showing (top) direct reduction of a ferric oxide mineral through contact with a bacterium and (below) the same process using an electron shuttle**

melanterite) in a process of bacterial oxidation that is greatly accelerated by the activities of other families of organisms (such as *Acidithiobacillus ferrooxidans* and *Leptospirillum*). For many years bacteria have been used in the mining industry to help in the breakdown of sulphide ores so as to release, and then recover, the metals.

Minerals and life

We now live in what geologists call the *Phanerozoic Era*, a stretch of time which began at the start of the *Cambrian Period* 541 million years ago. The word Phanerozoic is derived from the Greek meaning 'visible life' and is a reference to the first clear appearance of fossils which is in the rocks of this age. Although for the early geologists, there was no evidence of life in rocks older than the Cambrian (i.e. the *Precambrian*) many scientists, including Charles Darwin, argued that such evidence must exist but just had not been found. Trilobites and other relatively advanced creatures were well known from rocks of Cambrian age, even in Darwin's time, and must have evolved from ancestors of appreciably greater age. We now know that many organisms did exist before the Cambrian, but they were soft-bodied and were not preserved except under extremely rare conditions in rocks not known to the earlier workers.

The start of the Cambrian saw the evolutionary development of organisms with hard parts that were readily preserved in the fossil record. Then, around 520 million years ago, there was a development of life forms over a period of 10 million years of rapid evolution known as the 'Cambrian Explosion'. Much of this 'explosion' had to do with the evolutionary development of organisms with hard parts, with factors such as the development of predation leading to 'arms races' with some hard parts evolving as a defensive armour (shells) and others for attack (teeth). These hard parts were composed of minerals as they still are in the evolutionary descendants of those first shelly creatures or those with external skeletons. The minerals used by organisms in this way form some extraordinarily complex structures. These are *biominerals* and the remarkable processes by which living organisms produce them we call *biomineralization*. A relatively small number of minerals are formed in this way; mainly they are forms of calcium carbonate (calcite, aragonite), of calcium phosphate (apatite), and of silica, but they can be responsible for vast rock formations. Many of the limestones found through much of the geological column are the remains of great coral reefs similar in scale to the present-day Great Barrier Reef off the coast of Australia. Other limestones are largely composed of beds of shells: the ancestors of present day oysters.

A remarkable example of the capacity of living organisms to produce biominerals on a large scale is provided by *coccoliths*. The chalk rock familiar from the south of England, responsible for the 'White Cliffs of Dover', is 95–99 per cent composed of these microscopic plates of calcite which were used to form a protective armour around unicellular planktonic algae (*coccolithophorids*). As this example shows, as well as relatively large organisms, biomineralization is responsible for the hard parts of numerous microorganisms, many of which have delicate structures of great beauty. A good example is provided by the *radiolaria* (see Figure 17) which have skeletons of poorly crystalline ('opaline') silica. These free-floating single-celled organisms are found in the upper few hundred metres of the water column in the oceans, seaward of the continental slope. After death, their skeletal remains sink through the water column to contribute to the siliceous radiolarian oozes found on the ocean floors in equatorial latitudes. The fossilized remains of radiolaria are preserved in cherts and flints, which are rocks comprised of a microcrystalline variety of quartz. The key question in the study of biominerals is: 'How could these elaborate structures be produced by the organisms concerned'. Here, once again, the mineral and living worlds are intimately associated.

Two distinct mechanisms are proposed for biomineralization. In 'biologically induced biomineralization', the minerals form without apparent regulatory control, and as incidental by-products of the interactions between an organism and its environment. The minerals produced are very similar in their chemistries and crystal habits to those formed without any biological involvement. By contrast, there is 'biologically controlled biomineralization' in which the organism precipitates minerals that have specific physiological and structural roles. In this case, minerals can be formed within the organism even when the overall conditions in the solution surrounding it are not favourable for their formation.

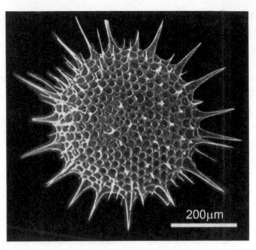

17. The 'opaline' silica skeleton of a radiolarian

Biologically controlled mineralization is a complex topic of great current research interest and one where we still have much to learn. For example, it seems that the formation of mineralized tissues is often closely linked to organic compounds that are concentrated in specialized regions ('compartments') within the organism during mineral formation. However, isolating those organics and assigning each one a particular function is problematic, particularly as isolating and removing any particular molecule may actually disrupt its function. Here, methods need to be developed for studying the relevant systems within the living organism (i.e. '*in vivo*'). The roles of amorphous and transient metastable mineral phases during biomineralization are important areas of investigation, as it appears that many organisms use amorphous precursors to mineralization. There is much to be learnt about how biomolecules promote the transition from amorphous to crystalline mineral whilst controlling the development of particular crystalline forms, habits, and faces.

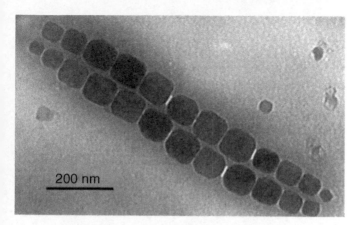

18. A double chain of magnetite crystals in a magnetotactic bacterium

Amongst the most remarkable examples of organisms producing a mineral to serve a specific function are the *magnetotactic bacteria*. Here, the bacterium concerned produces a chain of perfect crystals (see Figure 18), most commonly of magnetite (Fe_3O_4), which make use of the magnetic properties of that mineral. It seems that these organisms use the magnetite to become aligned in relation to the Earth's magnetic field and, therefore, in the most 'advantageous' position in relation to the environment at the sediment–water interface. The mineral greigite (Fe_3S_4) is found in other magnetotactic bacteria. It has very similar magnetic properties to magnetite but is stable in oxygen-poor, sulphur-rich sedimentary environments. The use of magnetic minerals for a specific function is also well known in a number of much larger organisms; birds in particular are known to use 'bio-magnetite' for navigation purposes.

Minerals and human health

Minerals, or rather the chemical elements from which they are made, are essential for humans but can also pose some of the greatest threats to health. Out of the 90 naturally occurring chemical elements on Earth, 22 are needed by a healthy human body and the great majority of those come originally from mineral sources. Deficiencies in the supply of such elements can lead to arrested development in children or pathological conditions in adults. Some of these requirements are widely appreciated, such as the need for calcium and phosphorous to form hydroxylapatite, $Ca_5(PO_4)_3(OH)$, the main mineral matter of bones and teeth. As well as the 22 essential chemical elements, more than 20 biominerals have been identified within the human body, with apatite being the most common. The rest include other calcium phosphates, along with carbonates and oxalates, again mostly of calcium and, in some cases, magnesium. Calcite, the mineral that dominates as a biomineral in so many organisms, does occur in the human body, but only in one very specialized site, in the maculae portion of the inner ear. Human biominerals can be divided into those which are an essential part of the body's systems, such as the material found in bones and teeth, and those which are unexpected and undesired ('pathological') mineral deposits. Although we talk of bone matter being dominantly hydroxylapatite, various substitutions for the calcium, phosphate, and hydroxyl ions in the structure leads to a much more complex chemistry, as suggested by a formula for 'bioapatite' of:

$$(Ca, Na, H_2O, \Upsilon)_{10}(PO_4, HPO_4, CO_3)_6(OH, F, Cl, H_2O, CO_3, O, \Upsilon)_2$$

where Υ signifies a site in the crystal that would normally contain an atom but is empty (or 'vacant'). Bioapatite may deposit in soft tissues or organs throughout the body, particularly causing problems in the cardiovascular system. Deposits of mineral matter can also build up in other parts of the body where they cause serious health problems. Examples of this are the 'stones' which can grow in human organs such as the kidneys or the gall bladder. In the kidneys, the stones are composed of the calcium oxalate minerals whewellite and weddellite and their formation has been linked to high levels of calcium in the urine. There are also examples of non-calcareous renal stones causing the pain in about 10 per cent of cases, and associated with the magnesium ammonium phosphate mineral struvite. These are related, in turn, to urinary infections caused by bacteria such as *Escherichia coli*. Pancreatic 'stones' or calcifications are mainly calcite, which blocks ducts leading to and away from this organ. Other pathogenic deposits include the crystals of monosodium urate monohydrate, which can form in the joint spaces of the extremities, causing gout, whereas crystalline deposits of three types (apatites, uric acids, and pyrophosphates) are associated with osteoarthritis.

Mineral derived materials can pose a serious threat to human health through their release into the food chain from poorly controlled mining or industrial waste disposal practices, or from abandoned industrial sites and mines. The materials involved can be toxic metals such as lead, cadmium, or mercury, or metalloids such as arsenic, antimony, or selenium. Our discussion in Chapter 4 of the release and transport of contaminants as part of the cycling of minerals at the Earth's surface is important here, and the extraordinary story of arsenic is described in Box 4. Three other special cases are worth discussing. One centres on the chemical property of radioactivity and the other two on the physical properties of particle size and shape. The health issues regarding radioactivity are associated with long-standing debates about the safety of nuclear power installations, the safe disposal of nuclear wastes, and the control of 'legacy' wastes from the former mining and processing of uranium ores.

Uranium and thorium are the important radioactive elements that occur in nature as minerals, and it is uranium that is mined for use in power generation, particularly as the oxide mineral uraninite (UO_2). Uranium has a complex chemistry that leads to a large number of minerals being formed by the alteration (especially the weathering) of uraninite. There are also the products of the radioactive decay of uranium from its ores such as the radioactive gas, radon. The damage to health caused by the exposure to

Box 4 Arsenic—the great poisoner

Throughout history, arsenic has been the poison of choice for murderers and assassins. Long before forensic analysis could detect even minute amounts of arsenic in a body, it featured in the power struggles of Imperial Rome and Renaissance Italy. One might imagine that arsenic poisoning leading to sickness and death is a fate no longer suffered by humans, but arsenic is currently responsible for what has been described as the 'greatest poisoning of a population in history'. In countries including Bangladesh, India (Bengal), and Vietnam, millions of people are affected through drinking water that contains arsenic concentrations above the World Health Organization 'acceptable' maximum level of 10 parts in a billion. The problem with arsenic is that it can accumulate in the body over many years. So where does this arsenic come from?

The story here is one of 'good intentions'. In attempts to reduce the occurrence of water-borne diseases, such as cholera, arising from the use of surface waters for human consumption in large areas of Bangladesh and Bengal, numerous wells were sunk to tap the water available in shallow aquifers. The water trapped in the aquifer sediments, consisting of fine grained mineral matter brought down to densely populated regions from the rapidly eroding Himalayas in great rivers like the Ganges, are the source of the arsenic. Arsenic was probably transported sorbed to surfaces or incorporated within very fine grained minerals such as iron oxides and hydroxides. Release of this arsenic appears to be associated with the microbial reduction of insoluble ferric iron to the soluble ferrous iron form, and of reduction of arsenic (from the As^{5+} ion to the more mobile As^{3+}). This microbial activity also requires a supply of organic carbon. Much recent research has been aimed at remediation. Strategies have included drilling into deeper aquifers, manipulating the chemistry and microbiology within the aquifer, or using the iron hydroxides formed on corroding metallic iron filings to treat the water and remove the arsenic at the well-head.

radioactive minerals or their breakdown products comes both from exposure to the radiation and ingestion of material into the lungs or via the food chain. However, the evaluation of the health risk is complicated by the wide variation in decay times of different radioactive isotopes. Some, like the isotopes of uranium, remain radioactive for millions of years, whereas others are no longer a threat after a few weeks or even days.

Although many dangerous radioactive elements such as plutonium, neptunium, and technetium occur only as products of the nuclear industries, they can interact with common minerals such as clays and iron oxides, becoming incorporated in their structures in ways that may lead either to their concentration or their dispersal. The nuclear metals participate in the cycling of minerals at the Earth's surface, particularly being involved in sorption and desorption processes, as described in Chapter 4.

There are two great concerns over human health in the context of the nuclear industries. First, there are the dangers posed by natural (or human-made, such as acts of terror) disasters, as happened at Fukushima, Japan, in 2011 when the power plant was severely damaged by a tsunami. Secondly, there is safe disposal of the most toxic of the nuclear wastes. It is sobering to recall that just one litre of such 'high-level' wastes in liquid form, if distributed evenly throughout the global population would inflict a lethal dose of radiation on every person on the planet. High-level wastes, defined as those in which the level of radioactivity is over a million times that regarded as acceptable for human exposure, contribute only one-tenth of 1 per cent of the volume of wastes generated, whilst contributing 95 per cent of the radioactivity.

The problems involved in the safe disposal of radioactive wastes involve minerals in two particular ways. The first is concerned with transformation of the waste into a form where it is immobilized, i.e. held in a solid material that is as resistant as possible to being broken down to release radioactive material into the environment.

This 'wasteform' needs to be resistant to attack from solutions of widely varying chemistries. Although synthetic glasses are commonly used in this way, an alternative strategy involves incorporating the radioactive elements into synthetic minerals of the kind known to accommodate such elements into their crystal structures. For example, Australian scientists have developed a synthetic rock (known as SYNROC) composed of a group of titanium-bearing minerals such as zirconolite ($CaZrTi_2O_7$) and perovskite ($CaTiO_4$). The former can act as a host for uranium, thorium, plutonium, and related metals, and the latter as host for radioactive strontium, sodium, and rare earth elements.

The second role minerals can play arises if the wasteform does break down and release its highly toxic components. Commonly it involves surrounding the wasteform, contained within stainless steel canisters, with mineral matter of the kind that would readily take up toxic contaminants from any escaping solutions. Candidates here include various clay minerals and zeolites; minerals known to have high capacities for sorption of metals from solution.

These first lines of defence against the escape of radioactive materials are called 'near field'; beyond that, the 'far field' is determined by the geology (and hence mineralogy) of any disposal site where wastes are buried. The minerals present will be critical for the choice of a suitable site and likely dominated by those impervious to fluids, or able to take up any escaping contaminants.

Knowledge of the behaviour of key radioactive species in all of the near- and far-field systems is essential information for assessing (and minimizing) the risks involved in any radioactive waste disposal strategy. This includes modelling the interactions between possible contaminants and any minerals they may encounter, so as to properly understand the potential mobility of any escaping contaminants.

The health impact of another group of minerals is entirely to do with the shape (more correctly termed 'habit') of the crystals involved. 'Asbestos' is not itself a mineral name but a term used to label a small number of minerals that occur with a very distinctive habit (called 'asbestiform'): individual crystals of the minerals grow as fibres because they are so elongated along one axis. In the extreme case, they resemble other fibrous materials such as cotton. Indeed, some mined asbestos has been woven so as to make heat and flame resistant fabrics used for gloves or suits.

Although there are six minerals that can be asbestiform, over 90 per cent of the asbestos mined is so-called 'white asbestos' and made up of a mineral named chrysotile, which is a hydrate magnesium silicate member of a clay mineral family called the serpentines. (An electron microscope image of a cross-section through a chrysotile fibre is shown in Figure 7c.) Other forms of asbestos which are also mined are fibrous members of the amphibole family referred to as 'blue asbestos' (crocidolite) and 'brown asbestos' (amosite).

The dangers to health posed by asbestos are mainly associated with ingestion into the lungs. There are three diseases associated with asbestos exposure: lung cancer, mesothelioma (cancer of the pleural and peritoneal membranes), and asbestosis, where the lung tissue becomes fibrous and ceases to function properly. The precise roles of the different kinds of asbestos (and other 'elongate' minerals) in causing lung disease have long been controversial. However, it does seem that the 'white asbestos' (crysotile) is less hazardous than the other (amphibole family) types. It also seems that the dangers posed by low-level exposure (for example, to the occupants where asbestos has been used in a building, such as a school) are negligible compared to the exposure suffered by miners and mineral processors in the asbestos industry.

The dangers posed by asbestos have been known for decades and legislation limiting human exposure is now widespread. One other example of a health impact that I will discuss has only come to public attention in the last few years. It concerns nanomaterials and nanotechnology, a topic we touched upon in Chapter 4. Various scare stories began to appear in the popular press once the industrial production of synthetic nanomaterials got under way. One story, inaccurately attributed to HRH Prince Charles, heir to the British throne, was that nanoscale robots would take over the world and transform everything into 'grey goo'. Such ideas may have a place in science fiction stories but show no signs whatever of becoming a reality. There is some concern, however, arising from the limited knowledge we have of the properties of many nanomaterials, and their possible impact on the environment including biological systems and, ultimately, on human health.

The health and safety issues mainly concern 'free' manufactured nanoparticles rather than those incorporated into new materials or devices. As discussed in Chapter 4, the properties of nanoparticles are commonly very different from their larger grain-size equivalents. So, the absence of any possible adverse effects cannot be inferred from the behaviour of macro-sized equivalents. The extremely small size of nanomaterials means that they are much more likely to be taken up by the human body. However, much remains to be learnt about their behaviour and interaction with biological processes in all living organisms. To properly assess the health hazards posed by manufactured nanoparticles, the whole lifecycle of these particles, from fabrication to ultimate disposal, needs to be evaluated. This includes, for example, their interactions with the surfaces of minerals when, if released into the environment, they become part of the cycling discussed in Chapter 4.

Chapter 6
Minerals as resources

The mineral resources taken from the Earth are now essential for human survival. The key roles they play in production of food and provision of water for domestic use and irrigation mean that they are truly 'vital'. Complex networks are in place to enable the production and distribution of food and water, as well as the vast range of material goods expected by people, particularly in the developed countries, ranging from cookers and cars, to televisions, computers, and mobile telephones. To take just one example, bread requires crops produced on farms using tractors and other farm machinery themselves made using different metals, paints, plastics, and rubber, as well as mineral fertilizers and, in some cases, irrigation systems. The harvested crops in turn need milling, transport to a bakery with all of its operating equipment, and transport of the baked bread to the shop or supermarket. The distribution network itself requires both the mineral raw materials to manufacture vehicles, and that needed to build the roads and bridges on which the vehicles travel. All of this requires the energy supplied by oil, gas, or some other source and, although the fossil fuels (coal, oil, gas) and many forms of alternative energy (wind, tidal, hydroelectric, solar, etc.) are not mineral based, nuclear power stations require the mining of uranium minerals as their energy source. Also, for all energy sources, mineral raw materials are still required for the equipment used in extraction and distribution.

Not surprisingly, the growth in the consumption of mineral resources in the recent past has been dramatic. A good illustration is provided by world production of iron ore, which in the year 1909 was 126 billion metric tons; by 1959 that had grown to 439 billion metric tons, and by 1999 to 1,020 billion metric tons. It then took just another decade to more than double to 2,240 billion metric tons by 2009. Iron, a raw material sometimes described as the 'backbone of industry', accounts for about 94 per cent of all the metals ever mined. Its importance has been strikingly captured in the words of Rudyard Kipling in his poem 'Cold Iron':

> Gold is for the mistress—silver for the maid,
> Copper for the craftsman, cunning at his trade.
> 'Good!' said the Baron, sitting in his hall,
> 'But iron—cold iron—is master of them all.'

Iron, particularly when alloyed with relatively small amounts of carbon or of other metals (nickel, cobalt, chromium, vanadium, molybdenum, tungsten, etc.) to give the numerous types of steel, is used in everything from the construction of buildings and bridges, to railway systems, ships, pipelines, motor vehicles, and domestic appliances. The growth in the mining and consumption of iron ore is driven by growing consumer demand, especially in countries like China where rapid industrialization is under way. Ultimately, of course, consumption is driven by a growing world population and the legitimate demands of people in poorer countries for the consumer goods enjoyed by others elsewhere.

A vivid impression of the extent of resource consumption in countries such as Britain and the USA is provided by considering the average 'consumption' of certain mineral raw materials by a single person. Most of these materials are not handled directly by us, of course, but used on our behalf in constructing roads, bridges, schools, hospitals, homes, vehicles, and in numerous other material goods. Taking the USA as an example, in an average lifespan of 75 years, an American will consume about 800 metric tons of non-fuel resources and 400 metric tons of fossil fuels. The former includes (as approximate figures) 400,000 kg of stone, 22,000 kg of cement, 11,000 kg of phosphate, 30,000 kg of iron and steel, 1,900 kg of aluminium, 800 kg of copper, 400 kg of zinc, and 4 kg of uranium. Put another way, the present population of the USA will consume a total of more than 330 billion metric tons of resources in their collective lifetimes.

Inevitably, this level of consumption has an impact on what most of us call 'the environment' and which Earth scientists now call the 'critical zone', meaning the topmost few metres of the solid Earth, the oceans and surface waters, and the lower atmosphere. The use of the word 'critical' is a reminder that this zone is the region of the planet on which we humans depend for survival. As long ago as the 1990s, scientists were suggesting that our human activities are moving more material around at the surface of our planet than all natural processes (such as weathering and erosion, volcanic eruptions, earthquakes) combined. The observation that human beings have now become the dominant 'geological agents' has prompted some geologists to suggest we change the name of the *epoch* in which we now live from *Holocene* (from the Greek meaning 'wholly recent') to *Anthropocene* (the age of humankind). It is not only the extraction of materials from the Earth that causes disruption but also, in certain cases, the use of the resource and disposal of associated wastes. Several categories of minerals are exploited, as I explain below.

Kinds of mineral resources

Ores are rocks containing minerals which, after removal from the Earth using open-pit or underground mining methods, are treated to extract metals from them. Commonly, the valuable ore minerals are sulphides or oxides of one or more metals and occur closely intergrown with other minerals of no value. Processing of the materials taken directly from the ground and which may contain less than 1 per cent of the metal being mined (and, in some cases, far less) initially involves crushing and milling to liberate the valuable minerals, some form of separation and processing to concentrate them, and then extraction of the metal from the mineral *concentrate* by smelting or a hydrometallurgical process (e.g. dissolving out the metal using a reagent such as a strong acid).

Only six metals have an average concentration in the rocks of the Earth's crust that is greater than one-tenth of 1 per cent by weight. These are magnesium, aluminium, silicon, titanium, manganese, and iron. For this reason they are known as the abundant metals. More than 30 other metals occur at lower concentrations (at parts per million, or even per billion, levels) and are termed scarce metals. These can be further separated into the precious metals (gold, silver, platinum), base metals such as copper, lead, tin, and zinc, and ferro-alloy metals such as chromium, nickel, tungsten, and molybdenum. The name 'base metal' was originally used by alchemists in the Middle Ages because these were the undesirable metals which the alchemists tried to change into gold or silver. Ferro-alloy metals are those used to alloy with iron to make special steels. Another group, known as the special metals, have a variety of applications, mostly in modern industries such as electronics. Examples include gallium, indium, niobium, and tantalum.

Industrial minerals are materials valued for particular physical or chemical properties they possess. As further discussed below, this includes minerals such as diamond, which is mostly valued for its hardness, along with other hard minerals used in abrasives such as garnet; asbestos is valued for its fibrous character, whereas clays such as kaolinite are a pure white, unreactive, fine grained powder used in paper-making, as well as for firing to make high quality ceramics. Other examples are fluorite, used as a flux for steelmaking, and barite for its high specific gravity employed in the 'heavy mud' lubricant used when drilling for oil.

Chemical minerals include those such as halite ('rock-salt'; NaCl), which is used not only as a food preservative and condiment but also as the raw material for the manufacture of chemicals such as hydrochloric acid or caustic soda, or the mineral apatite (calcium phosphate; $Ca_5(PO_4)_3(F,Cl,OH)_2$), which is used as a source of phosphate for fertilizers.

As well as the many individual minerals exploited by industry, of which those mentioned above are just a few examples, very large amounts of many types of rocks are used in construction as cut ('dressed') stone, as aggregates, and as the raw materials for cement and concrete. At the other extreme in terms of value and quantity, as discussed below, gemstones are often rare examples of common minerals of exceptional perfection in terms of brilliance or colour when cut or polished. In the section that follows, we say more about the industrial and chemical minerals that depend on mineral properties for their value. Ores are then the main focus of the sections on mineral deposit formation and on the role of plate tectonics in processes of formation.

Mineral properties and mineral resources

Diamonds to clays

Marilyn Monroe reminded us in song that 'diamonds are a girl's best friend'. True, gem diamonds are amongst the most highly valued materials of any sort, but about 80 per cent of the diamonds mined are unsuitable for use as gemstones, being too small or imperfect. The interest in these stones is because of another remarkable property of this mineral; it is by a long way the hardest naturally occurring substance known. Industrially it is used at the 'sharp end' of the cutting and grinding tools needed to fabricate numerous products from metals, alloys, and ceramics. As noted in Chapter 1, the reason for the great hardness of diamond, which is chemically just a form of pure carbon, lies in its crystal structure. In diamond, the atoms form a rigid framework in which every carbon atom is linked to another four carbon atoms at the corners of a tetrahedron (see Figure 1c). Equally remarkable is that graphite, another mineral composed only of carbon, is one of the softest minerals known. Again, this is because of its crystal structure (Figure 1b) in which layers of linked carbon atoms have only very weak chemical bonds between the layers, enabling them to 'slide' over one another. For this reason graphite is used in lubricants, particularly for the moving parts of certain machines and engines. (Historically, graphite was also mined on a small scale in the English Lake District and used to make pencils.)

The hardness of diamond is a useful property in its role as a gemstone, making it resistant to damage and wear. The optical properties of gems such as diamonds are what makes them objects of beauty when they are cut and polished to show these properties to advantage. In the case of diamond, it is the large variation of refractive index with wavelength of light ('dispersion') which produces the characteristic sparkle. The refractive index is a measure of the change in velocity of a beam of light on passing from the air into the mineral; a change that causes the beam of light to bend, and which is exploited when a gem diamond is cut so as to cause multiple reflections of the light entering the stone.

Diamonds are also valued for a range of technological applications. As well as being the hardest natural substance known, diamond has the highest thermal conductivity at room temperature, is highly resistant to attack from chemicals, is an excellent electrical insulator, is transparent to light and X-rays, and can be prepared as a superior semiconductor for electronic devices.

Other gems such as ruby and emerald are prized for their colour. In many cases, a coloured gemstone is just a variety of a common mineral which contains a small amount of a particular impurity substituting in the crystal structure. For example, ruby is a variety of the aluminium oxide mineral corundum (Al_2O_3) which is second only to diamond in hardness and, therefore, also used as an abrasive. Very small amounts of chromium replacing the aluminium atoms in corundum are responsible for the beautiful red colour of a ruby. Emeralds are a variety of the beryllium aluminium silicate mineral, beryl; their beautiful green colour is also attributed to the presence of small amounts of chromium and possibly vanadium.

At the other end of the scale from diamonds and other gemstones are the minerals produced in bulk and at low cost. As we have already noted, the use of clays in brickmaking, pottery, and other ceramics was amongst the first uses by mankind of any minerals. An important property of certain clays, appropriately called *fireclays*, is a resistance to very high temperatures which makes them ideal for lining the interiors of boilers and furnaces. Silicate minerals that originally formed at high temperatures, such as the olivines, are also heat resistant and used in the production of what are collectively termed 'refractories'.

Many minerals are used in industrial processes and never seen by the consumer. For example, fluorite (CaF_2) is used as a flux in steelmaking, and baryte ($BaSO_4$), because it is both of high density and inert, is used in the drilling 'mud' pumped down into the drill hole in oil exploration. In this case, it is the density of the baryte that helps to counter the pressures forcing the oil and gas towards the surface. A variety of widely available, relatively low cost, inert, white minerals are well suited to the least glamorous of the many jobs for which minerals are now used. These are the *fillers* used to provide bulk or other characteristics to a wide range of products; from rubber, plastics, and paints, to detergents, cosmetics, medicines, and toothpastes. The clay mineral kaolinite ($Al_4Si_4O_{10}(OH)_8$) is a good example of such a mineral. It is used in plastics, adhesives, paint, and particularly, as discussed in Chapter 4, in the paper industry. Kaolinite is particularly well suited to these applications because it is produced as an inert, fine-grained, pure white powder.

Clean-up and catalysis

The examples above focus largely on the physical properties of minerals, but chemical properties are also very important. Certain minerals and their synthetic equivalents play a key role in industry where substances are required to remove impurities from water or to catalyse important chemical reactions. A very good example of this application is provided by members of the silicate mineral group called 'zeolites'. These are water-containing aluminium silicates of sodium, potassium, and calcium (and to a lesser extent magnesium and barium). A typical example is the mineral chabazite, $(Ca,Na)_2Al_2Si_4O_{12}.6H_2O$. Zeolites have crystal structures in which the SiO_4 and AlO_4 tetrahedral units are joined together to form a framework in such a way as to leave large cavities containing water molecules. These cavities may be interconnected in one, two, or three directions. When zeolites are heated to about 350°C the water is driven off, leaving a crystal permeated with empty channel systems in up to three directions. These apertures are large enough to allow small atoms and molecules to pass through them, but not large ones. Zeolites can be used to extract impurities from water, whether those impurities are the calcium and magnesium removed to soften drinking water or highly toxic radioactive contaminants such as caesium-137, removed to clean up heavily contaminated water. They are also used to extract CO_2 and H_2S from natural gas, or sulphur and nitrogen oxides from smokestack gases. These pollutants are responsible for acid rain, and for the 'greenhouse gas' contamination of the atmosphere implicated in global warming.

Although the industrial applications of zeolites began with the natural (mineral) materials, more important today are the thousands of tons of synthetic zeolites made from the reaction of solutions of sodium hydroxide, sodium silicate, and sodium aluminates. Of the 206 zeolite structures known, only 40 are naturally occurring. These materials can be made with crystal structures tailored to specific applications, notably for use as catalysts. For example, synthetic zeolites with larger empty cavities are used to catalyse the breakdown of large organic molecules in the refining of oil. This 'catalytic cracking' is more efficient than heating the oil to cause its break-up and can be used to add hydrogen to the oil. This process of *hydrogenation* results in an increased yield of petrol (gasoline) from every barrel of oil. Although the use of zeolites in catalysis is big business, the greatest use of these materials today is in household washing powders.

How mineral deposits form

All the minerals and rocks that we exploit have been formed by geological processes. In the case of most rocks, these are major processes such as the emplacement in the crust and cooling of magmas, eruption and deposition of volcanic lavas and ashes, deposition of sediments from waters, whether as chemical precipitates or detrital grains, and the heating and compression of pre-existing rocks during metamorphism. Abundant quantities of the major rock types such as granites, limestones, sandstones, or marbles are available for use in the construction industries. In most cases, such rocks require only cutting up or crushing before being used.

The situation regarding the availability of 'common' rocks contrasts with other mineral resources, especially the 'scarce' metals. These require unusual processes in order to greatly increase the concentrations of the metal concerned above the average value in the Earth's crust, sometimes by a factor of many thousands. Although the concentration level required for a particular metal, in order that it be economically worth mining, depends on factors such as the current metal price, ease of extraction, geographical location of the deposit, and political environment in the country concerned, approximate levels can be suggested. Surprisingly perhaps, the highest figure is for mercury (~12,000 times) whereas for gold it is much lower (~250 times)—a reflection of the very high price commanded by the latter. Even the abundant metals need to be significantly enriched over average crustal values to reach concentration levels where mining can be undertaken at a profit. Iron has a crustal abundance of 5.6 per cent, so that a roughly tenfold enrichment is needed. For aluminium (crustal abundance 8.2 per cent), only around fourfold is needed.

In order to form mineral deposits, minerals can be concentrated in one of several ways. The first involves deposition from hot water ('*hydrothermal*') fluids, typically coming from a cooling mass of molten rock such as a granite magma which has forced its way ('intruded') into the upper part of the Earth's crust. As they cool, such magmas expel waters which contain substantial amounts of dissolved metals and other elements which are not components of the main granite minerals (feldspars, micas, and quartz) and so not taken up into those minerals. The heat from the magma may also cause these waters to mix by convection with other waters in the surrounding rocks. The hydrothermal fluids may initially be at temperatures of several hundred degrees, but kept from boiling by being under pressure—rather as in a 'pressure cooker'. These fluids will then force their way along joints in the cooling rock and along faults and fractures in the surrounding rocks, and with falling temperatures and pressures deposit minerals, including those rich in valuable metals, as *veins* (Figure 19). The veins (as, for example, in the mineral deposits of Cornwall in the UK) may contain a wide range of minerals exploited for the extraction of metals including copper, tin, tungsten, lead, zinc, and uranium. Nearer to the intrusion, the cooling mixture of fluid and the last portions of crystallizing melt may produce large crystals of both common minerals (feldspars, micas, quartz) and exotic (such as beryllium-bearing or lithium-bearing) minerals in what we call *pegmatites*.

Granite rocks can be associated with the melting of the continental crust; and rocks related to granites are associated with the melting of a subducted slab, formed when one tectonic plate dives under another at a destructive plate boundary. These magmas are

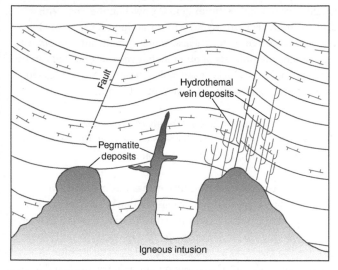

19. A simplified cross-section through the subsurface in a region above a shallow intrusion of an igneous rock such as a granite. Valuable minerals can be concentrated in pegmatite deposits and veins in the rocks surrounding the intrusion

responsible for chains of volcanoes such as those in the Andes of South America. It is in the root zones of such volcanoes that an important kind of deposit known as a *porphyry copper* deposit is found (and *porphyry molybdenum* deposits also occur in the same settings). In this case, the ore minerals such as the copper sulphide chalcopyrite or, in some cases, the molybdenum sulphide molybdenite are finely distributed through the rocks, forming a deposit that is a cylindrical mass of ore which is at, or just below, the surface. Such deposits average much less than 1 per cent copper and only became economically viable with the advent of very large-scale mining and processing methods in the mid 20th century. Very large amounts of ore can be removed from open-pit mines using modern methods requiring a comparatively small workforce (Figure 20). In some mines, more than 200,000 metric tons of ore are being extracted in a single day; an amount that would have taken those mining in Cornwall in the early 19th century many years to recover. One other important point about these mines is that their ores also contain small, but economically very significant amounts of other metals such as gold that can be recovered as by-products of the main operation. Such so-called 'sweeteners' can ensure the profitability of a mine.

A major advance in our understanding of hydrothermal fluids and their role in forming mineral deposits came from discoveries made in the 1970s in the course of ocean exploration.

The most spectacular of these discoveries was made in the vicinities of mid-ocean ridges, where molten basaltic rock rises from the mantle and spreads outwards forming new ocean floor. Here, hydrothermal fluids are generated as ocean water circulates through the hot rocks at and near the ridges. These fluids, at temperatures of ~350°C, rise rapidly through fractures and form jet-like eruptions called *black smokers* when they mix with cold ocean water, causing the dissolved material they contain to precipitate. In fact, the plumes of what look like black smoke are

20. **A modern large open-pit mining operation**

fine particles of iron, copper, zinc, and lead sulphide minerals (pyrite, chalcopyrite, sphalerite, galena). Processes such as those associated with the black smokers have occurred in the geological past and produced rich ore deposits, in particular, those known as *volcanogenic massive sulphides*.

Away from the ridges, sitting on the floor of the deep ocean, is another mineral resource, one of more uncertain origins. Here we find pea-to-grapefruit sized spherical objects called *manganese nodules*. Internally the nodules have an onion-like layering. These objects probably formed very slowly, and involved much lower temperature fluids, with microbial activity possibly playing a role. As well as iron and manganese oxide minerals, they contain valuable concentrations of nickel and cobalt. Although very challenging, the mining of nodules from the deep ocean is likely to be technically possible. But who owns the mineral resources of the oceans? It will probably come as no surprise to learn that there is no international agreement about the answer to that question, despite decades of discussions.

A second way in which mineral deposits can form is associated with the cooling, crystallization, and solidification of a magma, this time a magma coming originally from the upper mantle of the Earth, and formed by the partial melting of mantle material (possibly due to a local build-up of heat from decay of radioactive minerals). In this case, the magma is comparable in overall composition to the basalts that make up the ocean floor, but it has not reached the surface; instead, it has reached a position in the crust where it will slowly crystallize. This slow cooling allows *fractional crystallization* (see Chapter 3) to take place, in which different minerals crystallize from the melt and settle to the bottom of the magma chamber in succession, so as to form a layered body. In this way, particular minerals become concentrated in particular layers. Notably, chromite ($FeCr_2O_4$), the mineral that provides world supplies of chromium, is concentrated in certain layers by this process. Also found concentrated in some layers may be platinum and other platinum group element (palladium, osmium, iridium, rhenium, and ruthenium) minerals. The world's largest deposits of chromium and the platinum metals formed in this way are found in the Bushveld Complex in the Republic of South Africa.

Whereas the settling of crystals from a molten body of magma to form layers is one form of *precipitation*, more familiar perhaps is the equivalent process which occurs when seawater in an enclosed water body such as a saline lake or sea (like the Dead Sea on the Israel–Jordan border) evaporates, depositing layers of salts. The most important of the salts making up these *evaporite* deposits are the minerals gypsum ($CaSO_4 \cdot 2H_2O$), carnallite ($KCl \cdot MgCl_2 \cdot 6H_2O$), and halite (rock-salt; $NaCl$) itself. Other economically important minerals precipitated from seawater under appropriate conditions include apatite, $Ca_5(PO_4)_3(OH,F)$, which is a major source of the phosphorous needed for fertilizers. Probably the most important resource now mined almost entirely from deposits formed by sedimentary processes are those of iron; the *ironstones* and in particular the *banded iron formations* (BIFs). We are still not sure how BIFs formed. They show very regular layering or irregular banding on a fine scale (millimetres to centimetres) with layers of nearly pure silica (chert) and nearly pure iron oxide minerals: haematite (Fe_2O_3) and magnetite (Fe_3O_4). An important feature of BIFs is that they formed during a period between 2.6 and 1.8 billion years ago when the Earth's atmosphere was probably very different, with little free oxygen and perhaps more carbon dioxide than today. Under these atmospheric conditions, iron could have been transported much greater distances dissolved in water than would be possible today. The BIFs seem to have formed in broad basins into which soluble iron may have been introduced and then precipitated. The cause of the repetitive precipitation of alternating layers of iron oxides and silica has been a point of much debate, with suggestions including annual climatic changes, cyclic periods of evaporation, episodic volcanism, and even microbial activity.

The formation of evaporites or ironstones such as BIFs involves a chemical precipitation from water of the minerals concerned. Water can also play a role in concentrating minerals when it is flowing and the minerals are already present as detrital grains. To be concentrated in this way, the minerals need to be very resistant to corrosion and denser ('heavier') than the average minerals that make up common rocks. The concentration process is like that employed by the old style prospector panning for gold (see Box 5). The gold pan is a large, flat-bottomed metal bowl into which the prospector scoops up water and sediment from a stream. By employing a circular 'swirling' motion, the less dense mineral grains are winnowed out and discarded over the edge of the pan until only the densest grains are left. Amongst these the prospector hopes to see the glint of a few grains of gold, washed downstream from the primary source of the metal, the 'mother lode'. When nature concentrates gold or other minerals by the action of flowing water the result is a *placer* deposit. Besides gold, other metals extracted by the mining of placer deposits include tin as cassiterite (SnO_2), and titanium as rutile (TiO_2) or ilmenite ($FeTiO_3$). Other valuable minerals recovered from placers include diamonds, rubies, and sapphires.

A final process to mention is that responsible for the formation of *residual* mineral deposits. All rocks and minerals exposed at the Earth's surface are broken down by the weathering action of rain, wind, frost, and the action of microbes including fungi. The materials in a rock which are least stable to such weathering may be transported away, or there may be some redistribution of the material making up the fresh rock. Such processes may result in the selective concentration of valuable minerals. An example is illustrated in Figure 21. Here, peridotite, a rock of a type originating from the upper mantle and rich in minerals such as olivine, $(Mg, Fe)_2SiO_4$, and pyroxene, $(Ca,Mg)SiO_3$, has been exposed at the Earth's surface and is weathering. The fresh olivines and pyroxenes in this rock contain small amounts of nickel (~1 per cent) which is released when these minerals, which are unstable when exposed to the atmosphere, break down. As seen in the depth profile in Figure 21, the heavily weathered zone nearest to the surface is altered to a *laterite*. This iron-rich,

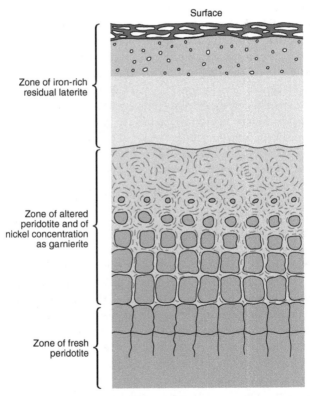

21. A cross-section of the shallow subsurface where the minerals of a peridotite rock are being weathered, resulting in the concentration of mineable quantities of nickel

bright orange-red residual material largely made up from iron hydroxide minerals such as goethite (FeOOH) is characteristic of the weathering of rocks such as peridotites in warm tropical regions of high rainfall. Below the zone of residual laterites, and above the fresh peridotite, is a zone of altered peridotite. In this zone, the nickel is concentrated after its release from the olivines and pyroxenes, and is redeposited in the form of a nickel-rich clay mineral called garnierite.

Although these residual deposits make an important contribution to the world supplies of nickel, the most important such deposits are those of aluminium. The breakdown product in this case is an aluminous laterite called *bauxite* which is a mixture of several aluminium hydroxide minerals. The aluminium was originally a component of silicate minerals such as feldspars, but under certain weathering conditions, the silica and other components are dissolved away leaving rich ores (sometimes reaching aluminium concentrations of around 40 per cent).

Box 5 The strange story of gold

We have prized gold since the emergence of the first civilizations. As a metal it is soft and malleable, and also extremely resistant to corrosion, and it occurs as the 'native' element making it easy to extract and exploit. Gold can be found as veins where it has been deposited from hydrothermal ('hot water') solutions in fractures in the vicinity of granites and related rocks. When veins like these are weathered away, the gold may be transported in flowing water in streams and rivers. Being resistant to corrosion and very dense (around 19 times the density of water) it may then become concentrated in parts of a stream where the flow is slowed, such as the inside of a meander loop. Processes like these have had a key role in forming the world's greatest gold deposits in the Witwatersrand Basin in South Africa. These deposits were discovered in 1886 and rapidly became the main producers in the

Box 5 (Continued)

world, a position they still occupy today. However, during the past 50 years, developments in 'open-pit' mining and in ore processing methods have seen increasing competition from very large 'low grade' deposits where the gold is a minor component of the extraction operation. For example, the *porphyry copper* deposits associated with destructive plate boundaries can contain significant gold.

The lure of gold has long been a part of the human condition, notably recorded in the great gold 'rushes' like the one in California in 1849. Few who set off to the goldfields made a decent living, let alone a fortune, from mining; arguably those who did best were people like Levi Strauss, a tailor who made his money selling denim trousers ('jeans') to the prospectors. The total amount of gold ever mined is remarkably small, estimated to be about enough to fill the space occupied by less than four Olympic sized swimming pools. It is also the case that nearly all of the gold ever mined is still in circulation. The gold ring worn by a reader of this book could well contain atoms of gold that once adorned a famous queen like Cleopatra. Perhaps the strangest aspect of the story of gold is the role it still plays in currency and global finance. Although no longer used in coinage, a very large proportion of the gold we have is buried, not back in the ground but in secure vaults, by governments as a back-up to their currencies.

Plate tectonics and mineral resources

An important question, given the processes of formation of mineral deposits we have described, is: 'Do these processes relate to the plate tectonic cycle and, if so, how?'

The answer is that many, but certainly not all of these processes can be understood as part of the cycles of formation and destruction of lithospheric plates that we have discussed in Chapter 3. We can consider some examples of mineral deposits mined for metals, examples where plate tectonics provides a global context for understanding what geologists call 'ore genesis'.

Figure 22 is a much simplified 'slice' through part of the lithosphere and uppermost mantle, showing both constructive and destructive plate margins with an indication of the types of deposits found in a particular setting, and the metals extracted from those deposits. In line with the above account of mineral deposit formation, at a constructive margin (mid-ocean ridge), upwelling of magma brings molten rock in contact with seawater; this water is therefore heated and circulates so as to concentrate metals in hydrothermal fluids which are vented into the ocean bottom waters as the metal sulphide-carrying black smokers. This material can accumulate to form rich deposits of copper and zinc ores: the *massive sulphide* deposits. The ore minerals found in such deposits are dominantly chalcopyrite ($CuFeS_2$) and sphalerite (ZnS).

Moving away from the ridge to the deep ocean floor we have the manganese nodules, sources of cobalt and nickel as well as manganese (dominantly as manganese oxide minerals), before reaching the subduction zone where ocean crust is dragged down into the mantle. Not surprisingly, the region of the crust where a slab of ocean floor crust is dragged down is one where massive slabs of rock can be thrust up against one another along with modern sediments in a great jumble of rocks called a *mélange*. In this environment, some of the rocks and mineral deposits that we find actually originate at depth beneath the seafloor in the vicinity of the mid-ocean ridges, and were transported with the ocean crust as it spread. Important examples include deposits of chromium (as chromite; $FeCr_2O_4$) which originated by fractional crystallization from magmas trapped beneath the surface. They were transported with the spreading ocean floor before being thrust up against the rocks of a continental mass as a stranded piece of ancient ocean floor or *ophiolite* (from the greek 'ophis' meaning 'snake', and a reference to the snakeskin-like appearance of the rocks involved).

Figure 22 emphasizes just some of the complexities that can arise at a destructive plate boundary. Here, magmas generated by the partial melting of the down-going slab can form a magmatic ('island') arc. In these settings, the interactions between plates can also put the crust under tension and cause stretching of the crust to form ('back arc' and 'outer arc') basins. These basins can be the sites of mineral deposits which may be in layers parallel to the surrounding strata (*stratabound*) or may be irregular masses (*massive sulphides*). These ores are important sources of base metals including copper, lead, and zinc. As noted above, the root zones of the volcanoes of the magmatic arc is where circulating fluids concentrate metals as *porphyry copper* or *porphyry molybdenum* deposits. The main ore minerals in porphyry copper deposits are pyrite, chalcopyrite, and bornite (Cu_5FeS_4) and in porphyry molybdenum deposits, molybdenite (MoS_2) is the source of that metal. Minor amounts of native gold are commonly found as 'sweeteners' in these deposits.

Finally in Figure 22, the crustal rocks of the continents host plutons of granite and related rocks where veins (see Figure 19) may carry ores which are sources of tin (as cassiterite; SnO_2) and tungsten (as wolframite; $(Fe,Mn)WO_4$) along with other metals including copper, uranium, and bismuth (as bismuthinite; Bi_2S_3). Where granitic magmas have reacted with the rocks into which they have been intruded (the process of 'contact metamorphism'; see Chapter 3), a wide range of deposits may be formed, variously providing sources of tin, tungsten, molybdenum, copper, lead, and zinc.

The role of the plate tectonic cycle in the formation of certain mineral deposits has significance in the search for new deposits. It enables the geologist engaged in mineral exploration to focus attention on appropriate areas of the Earth's crust; areas where

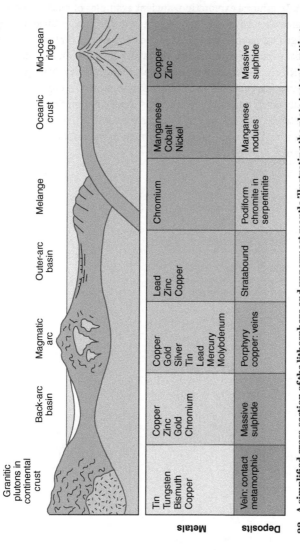

	Granitic plutons in continental crust	Back-arc basin	Magmatic arc	Outer-arc basin	Melange	Oceanic crust	Mid-ocean ridge
Metals	Tin Tungsten Bismuth Copper	Copper Zinc Gold Chromium	Copper Gold Silver Tin Lead Mercury Molybdenum	Lead Zinc Copper	Chromium	Manganese Cobalt Nickel	Copper Zinc
Deposits	Vein: contact metamorphic	Massive sulphide	Porphyry copper: veins	Stratabound	Podiform chromite in serpentinite	Manganese nodules	Massive sulphide

22. A simplified cross-section of the lithosphere and uppermost mantle illustrating the plate tectonic settings of a number of kinds of mineral deposits

the geology is appropriate for the formation of the deposits of interest. In mineral exploration, ancient examples of plate tectonic environments such as subduction zones need to be identified by careful geological mapping and the use of other exploration methods. In response to the age-old saying 'how do you find an elephant?' comes the answer 'look in elephant country'. Mineral exploration is a challenging activity because mineable deposits form only rarely and as the result of a special combination of circumstances. We will return to issues concerning the search for mineral resources, and the scarcity of such resources, in the final chapter of this book.

Chapter 7
Minerals past, present, and future

Mineral evolution

According to current thinking, when the Universe began the only elements that existed were hydrogen, helium, and traces of lithium. There were no minerals. Millions of years after the Big Bang came the formation of the first stars within which all the other elements were made via nuclear reactions. Then as giant stars exploded in the first supernovae, it is proposed that the first crystalline matter, the first minerals in fact, formed in the cooling stellar envelopes. This would have been a carbon-rich environment, with diamond and graphite likely to have been the most abundant crystalline phases, together with lesser amounts of carbides, nitrides, oxides, and magnesium silicates. How did we get from this dozen or so primeval minerals to the more than 4,000 minerals known today? Strange as it may seem, it has only been in the last few years that this question has been systematically addressed through the concept of 'mineral evolution' as proposed by a group of mineralogists in the USA led by Bob Hazen of the Carnegie Institution of Washington.

In what we might call the 'theory of mineral evolution', it has been proposed that the Earth (itself accreted from a cloud containing the detritus of various supernovae and circling our early Sun) has evolved through three *eras* which can themselves be further subdivided to give a total of ten *stages*, with each adding to the total inventory of mineral species. The three eras have been named as the eras of *planetary accretion* (Stages 1 and 2), *crust and mantle reworking* (Stages 3, 4, and 5), and *biologically mediated mineralogy* (Stages 6, 7, 8, 9, and 10).

During Stage 1, approximately 4.6 billion years ago, about 60 mineral species formed as condensates from the Solar Nebula. These were the essential planet forming materials such as iron–nickel alloys, sulphides, phosphides, and the oxides and silicates which are stable to high temperatures. They would soon have clumped together to form the small planetary bodies known as 'planetismals'. Stage 2 would involve some of these bodies becoming large enough to partially melt and, as a result, for their minerals to undergo alteration processes forming new minerals. This would have led to an enlarged inventory of about 250 minerals; those minerals found today in the diverse types of meteorites that fall to Earth.

Stage 3, the beginning of the crust and mantle reworking era (*c*.4.0–4.5 billion years ago) would have involved igneous processes such as volcanism, planetary degassing, fractional crystallization, metamorphism, and the beginnings of plate tectonics with its large-scale water–rock interactions. As discussed in Chapter 3, the formation of the different types of igneous rocks was described as an 'evolution' by N. L. Bowen in his classic work. All of these processes, which would have included a major role for volatiles such as water and carbon dioxide, might have brought the total number of mineral species to around 500.

Stage 4 (at an estimated time 3.5–4.0 billion years ago) requires the planet to have sufficient inner heat to melt its original basalt rock crust so as to form granite-type rocks and even pegmatites (coarsely crystalline, late-forming rocks produced by fractional crystallization and containing rare elements such as lithium, beryllium, boron, niobium, tantalum, and uranium). At this point the mineral inventory would total about 1,000 species. The final stage (Stage 5) of this second era is associated with fully developed plate tectonics. Volcanism associated with spreading centres and with subduction zones would have led to very large-scale water–rock interactions creating new minerals. Furthermore, there would have been the uplift associated with mountain-building leading to exposure at the Earth's surface of the minerals in what had been deeply buried rocks. In this way our total number of minerals reaches around 1,500. Much of what we have described up to this point appears to hold true for the other rocky planets that we have begun to explore in our Solar System, but on Earth this number of 1,500 falls very far short of the total of around 4,400. So what makes the Earth so different? The answer is biologically mediated mineralogy.

During the first designated stage of the third era (Stage 6; 2.5–3.9 billion years ago), the Earth's atmosphere was still without oxygen, and it is suggested that few new minerals were added at this time. The situation changed dramatically as a result of what is termed the 'Great Oxidation Event' (GOE) caused by the development of photosynthesis by certain bacteria and therefore of an oxygen-rich atmosphere (Stage 7; 1.9–2.5 billion years ago). This probably began about 2.4 billion years ago, at which time atmospheric oxygen may have risen to a little more than 1 per cent of present-day levels. More than half (about 2,500) of all known minerals are produced when other minerals are exposed to oxygen and oxygen-carrying waters at the surface of the Earth. Examples include the numerous clay minerals formed by the weathering of silicates, or the sulphates and hydroxides formed by alteration of sulphides. These would not have developed in an oxygen-poor environment, that is before the GOE changed the face of our planet.

The billion years or so following this rapid growth in the number of mineral species (Stage 8; 1.0–1.9 billion years ago) has been called the 'Intermediate Ocean' and is associated with relatively little activity that would give rise to new minerals. The oxygen-rich zone in the uppermost waters of the oceans may have deepened, but there is no evidence of dramatic developments. The ninth stage (0.542–1.0 billion years ago) is distinguished by at least two global glaciations referred to as 'Snowball Earth' episodes. It is still debated as to whether ice ever completely covered the Earth, but it certainly dominated for periods of more than 10 million years. Beneath the ice, volcanic activity continued and would have contributed to the mineral inventory, as would activity during interglacial periods with a likely rapid increase in the generation of new clay minerals.

The final stage (Stage 10) embraces the whole of what geologists call the Phanerozoic. As discussed in Chapter 5, the Phanerozoic spans from 541 million years ago until the present day. The clear fossil evidence for living forms which characterizes the start of the Phanerozoic was particularly associated with mineral skeletons made of carbonate, phosphate, and silica mineral matter. These were easily preserved, providing a rich and varied 'fossil record'. Later, when life became established on land as well as in the seas, the rise of land plants would have led to the first soils; these conditions contributed further members to the families of clay minerals, completing the story of mineral evolution.

The emergence of life

Any parallels between mineral evolution and the evolution of life forms can only be taken so far. We are not seeing the driving force that is provided by natural selection in Darwinian evolution, or the common thread of an inheritable molecule (DNA). And whereas around 98 per cent of the living organisms that have ever existed are now extinct, minerals must be very rarely, if ever, lost from our list of 4,000 plus species. There is, however, one period in Earth history where the biological and mineralogical may truly have converged. The evidence that minerals played a key role in the emergence of life on Earth is very strong. Just a few of the ideas put forward in this complex and diverse field of research will convey some aspects of current thinking.

The complex series of steps leading to the emergence of life must have started with the formation of the organic molecules required as life's building blocks, such as amino acids, lipids, and sugars. In what was probably the most significant laboratory experiment ever conducted in the quest for understanding life's beginnings, Chicago University student Stanley Miller and his supervisor Harold Urey showed that such molecules could be synthesized when a simple mixture of gases is subjected to electrical sparks simulating a lightning strike. The result was a so-called 'primordial soup' rich in the essential components needed for life. But then the question became: 'How did the simple molecules formed in this way, and greatly diluted in Earth's earliest oceans, come together to build more complex molecules and eventually form self-replicating entities? Some sorts of templates would have been needed and the perfect answer is provided by the surfaces of minerals. For example, the surfaces of common minerals such as quartz and calcite have been shown experimentally to select and concentrate specific biologically important amino acids, the building blocks of proteins.

A role for minerals as catalysts for biochemical reactions and templates in the emergence of complex biomolecules is now widely accepted. However, many different routes have been proposed for the emergence of the first living organisms; significantly, almost all hypotheses have major roles for minerals. Some routes, such as proposed by Joe Smith, Ian Parsons, and co-workers, involve minerals that have biomolecule-sized cavities in their crystal structures or weathered surfaces; they could have acted both as templates and as catalysts for biochemical reactions. Good examples of such minerals are the zeolites, which are framework silicates with cavities and tunnels in their crystal structures that can accommodate organic molecules and promote reactions between them, as shown by their widespread use as catalysts in industries such as oil refining (see Chapter 6). Others involve clay minerals such as montmorillonite in the formation of the first self-replicating genetic molecules. The experimentally demonstrated capacity of montmorillonite to catalyse the construction of more complex, longer chain biomolecules from RNA (ribonucleic acid, which some believe to have been a necessary precursor of DNA), is strongly suggestive of a key role for minerals. Probably the most extreme case where a critical role is envisaged for clay minerals has been advocated by the Glasgow chemist, Graham Cairns-Smith. He has argued that the driving force for the transition from geochemistry (minerals) to biochemistry (biomolecules) was a form of 'natural selection' operating initially just on inorganic materials. In effect, it is suggested that the varying stacking sequences found in the layering of clay minerals could retain and transmit information in a similar, but very much cruder, fashion to that now transmitted by DNA.

Another direction taken in the 'emergence debate' focuses attention on a very different group of minerals, the metal sulphides; this approach took its initial inspiration from the hydrothermal vents discovered on the ocean floor in the proximity of mid-ocean ridges. Here, hot fluids (~350°C) expelled into the ocean waters as 'black smokers' release a stream of metal sulphide particles onto the ocean floor and build 'chimneys' from the deposited sulphides. Both micro- and macro-organisms utilize chemical energy available in these environments for their metabolisms. Advocates of a central role for sulphide minerals also point to the central role played by transition metal sulphide clusters in key microbial enzymes and, hence, in the metabolic chemistry of many microorganisms. Two notable hypotheses are associated with the names of the German chemist Gunter Wachtershauser and an originally Glasgow-based scientist, Mike Russell. Both propose that life emerged through abiotic chemical reactions catalysed by metal sulphide minerals and involving fluids coming from depth in the Earth, probably emerging on the floor of an early ocean. Both also propose that the first organism was a so-called *autotroph*, a life form that can manufacture its own biomolecules from small molecules.

Wachtershauser suggested that pyrite (FeS_2) formation from the reaction of iron monosulphide (FeS) with hydrogen sulphide (H_2S) provided an energy source for the first life forms and a route to forming key organic molecules such as formic acid. He also suggested that pyrite could act as a catalyst for a wide range of reactions that would produce simple biomolecules. Russell and co-workers envisage hydrothermal fluids mixing with ocean water on the sea floor, rapidly precipitating iron sulphide (FeS, the mineral mackinawite) to form bubbles; these bubbles would serve as primitive membranes which could control and catalyse essential biomolecule-generating chemical reactions.

Many experiments have been conducted over the past two decades that demonstrate the capacity of minerals to catalyse reactions that must have been important for the eventual construction of the biomolecules needed for living organisms. There are also tantalizing clues regarding the transition from non-living to living worlds in the ways in which minerals can act as templates for the construction of complex molecules.

Resources for the future

Theories of evolution, whether it is evolution of minerals, or of biochemical systems leading to the first living organisms, are intellectually challenging, but generally focus our attention on the past. For the present and the near future, our greatest 'mineralogical' concerns are about practical issues centred on minerals as resources. They are also about the impact on our environment of the extraction, processing, and utilization of mineral resources, and disposal of the associated waste products.

As we have outlined in earlier chapters of this book, minerals are the sources of the majority of our material needs apart from food, water, and certain forms of energy, and even those rely heavily on mineral-derived products for their production and utilization. Great demands will be placed on mineral supplies in the coming years, so a key question is 'Will the Earth be able to provide our future mineral needs?' The answer to this question depends on the particular resources being considered. Metals are a good case in point. Iron and aluminium have become the principal metals for the transport and construction industries and for manufacturing many other essential products. Reserves of these two metals are vast, and technologies for their extraction are well developed, as are systems for their recycling. The other (geologically) abundant metals—magnesium, manganese, silicon, and titanium—also have large reserves and established patterns of use which look set to continue.

Twenty-five years ago, the reserves of the (geologically) scarce metals were thought to be running out. However, advances in exploration methods and improvements in recovery of metals from the mined ores and in recycling have greatly improved the situation in recent years. The precious metals are likely to continue to hold their allure in jewellery and as investments, and see a growth in technological applications.

New alloys may add to the demand for ferro-alloy metals. The base metals may, in some cases such as the toxic metals lead and mercury, see declining use because of environmental concerns. Others such as copper may find new uses. However, it is in the case of the special metals where demand could outstrip supply. Metals such as niobium, tantalum, germanium, gallium, indium, beryllium, and the rare earth elements (REE) are of crucial importance for modern industries, especially electronics. Niobium (also called columbium), as well as being used in high-strength alloys, is used for special types of magnets (superconducting magnets). Tantalum is used in electronic components which are essential for mobile telephones, DVD players, and computers. REE are to be found in every car, computer, smartphone, energy efficient fluorescent lamp, and colour television.

Because of their roles in advanced technologies, the special metals can also be strategically important. This is illustrated by the REE, for which the world supply has been effectively limited to several mining districts in China. This dependence was already a cause for concern when, a decade ago, the Chinese government introduced export quotas restricting the supply of REE outside China. Initially, supply matched demand but in 2010, a reduction in quotas of 40 per cent over 2009 levels was introduced which also led to very large price increases (over 1,500 per cent) in just a few months. Not surprisingly, these developments have prompted worldwide exploration efforts in attempts to find supplies of REE from outside China. Controversially, rich deposits appear to be located in unspoilt areas such as South Greenland. Another, but contrasting example of the impact that having deposits of a strategic metal can inflict on a country is provided by the Democratic Republic of Congo. Here, the tantalum that is mined (as ores referred to locally as 'coltan', a name derived from the minerals columbite-tantalite) and exported to European and American markets, is cited by experts as a key factor in financing civil wars in that region. In the mining itself, child labour is commonly used in appalling working conditions.

What of mineral resources other than metals? The minerals used for fertilizer, chemical, and building materials exist in large quantities and reserves will be adequate for many years. Marine potassium and phosphate deposits are extensive and sufficient to meet world needs far into the 21st century. The oceans and saline lakes are likely to serve increasingly as sources of chemical minerals. Building materials will always be available, even if local shortages may arise, and much the same can be said of most industrial minerals, even though some, such as the asbestos minerals, may be replaced by more environmentally friendly alternatives. Energy sources are always going to be problematic, given the great and growing energy demands of modern societies. Only one source is directly dependent upon minerals for its energy and that is nuclear power. Although a certain amount of secrecy surrounds the mining of uranium ores, there is no suggestion that serious shortages are about to arise, and substantial deposits are known on all the continents.

Returning to the question of whether the Earth's mineral resources are enough to meet our future needs, in many cases the answer is clearly 'yes'. In cases such as certain of the scarce metals, exploration in more hostile terrains may be a way forward. In particular, exploration and exploitation of deeper parts of the continental crust may be a way to supply our needs. Our knowledge of the subsurface at depths greater than ~1,500 feet is extremely limited; new techniques and programmes aimed at remotely mapping these deeper regions should reveal new deposits, but these will also pose new challenges for mining methods. Another way of addressing this question is to consider potential sources of metals where the amounts (percentages) of the metal are lower than those currently being economically mined. This can be illustrated by the example of copper. Figure 23 is a hypothetical depiction of copper production in the centuries ahead. One axis of this graph shows the percentage of copper in the mined ore, mining beginning with the exploitation of sulphide mineral ores with several per cent copper down to about one-tenth of a per cent copper content. After these have been exhausted, attention could turn to mining of deep ocean manganese nodules and metal-rich deep ocean muds containing smaller amounts of the metal. As a last resort, common rocks such as basalts, containing very low concentrations of copper, could be mined. However, the decreasing economic viability of following the pathway shown by the arrows on this diagram is emphasized by the information on the other axis showing the estimated energy required to recover a pound (0.454 kg) of copper from these different sources. As the percentage of copper (the 'grade')

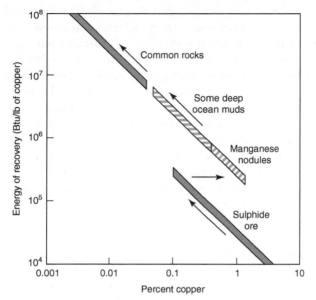

23. **A hypothetical illustration of sources of copper for future exploitation; the percentages of copper available from different sources are plotted against rising costs of extraction expressed in terms of Btu (British Thermal Units) per pound (lb) weight**

decreases, the cost of extraction in energy terms (and financially) increases. In addition, recovering copper from sulphide minerals or oxides (as in nodules) is much easier than recovering it from solid solution in rock-forming silicate minerals. The latter requires crossing what has been called a 'mineralogical barrier', with new extraction technologies being needed. What is shown in Figure 23 is conjecture of course, but it highlights the point that recovering scarce metals will be more difficult and expensive in the future.

An epilogue

Our introduction to minerals has taken us on a journey in space and time; from deep within the Earth to the outer regions of the

Solar System and beyond, and from the beginning of time through prehistory and history to the present. Minerals have always been, and always will be, central to our efforts to understand our planet, its past and its possible futures. Today, we are at the threshold of a new understanding of the Earth's surface processes which integrates the mineralogical, geochemical, and biological realms at the molecular scale. The emergent field of 'molecular environmental science' should provide new insights into the way the planet 'works' that could be comparable to the revolutionary advances seen in human biology associated with the genetic code. These ideas are associated with what is now being called 'Earth System Science'. However, one of the great founders of the science of geology, James Hutton, was the first to see the Earth as a 'system'. In 1785 he wrote:

> A theory is thus formed, with regard to a mineral system. In this system, hard and solid bodies are to be formed from soft bodies, from loose or incoherent material, collected together at the bottom of the sea; and the bottom of the ocean is to be made to change its place with relation to the centre of the earth, to be formed into land above the level of the sea, and to become a country fertile and inhabited.

Regarding the practical applications of our knowledge of minerals and of Earth Systems, the health and well-being of humankind are linked to minerals as both sources of essential nutrients or as potential poisons, and as the providers of the raw materials which are vital for our survival. Our dependence on minerals as resources can only increase in importance as world population grows from its present nearly 7 billion to the more than 9 billion predicted by the year 2050. For many of the challenges facing humanity in the coming decades, minerals will play a central role. Truly it can be said that 'minerals matter'.

Further reading

Elements: An International Magazine of Mineralogy, Geochemistry and Petrology (2005–present; Mineralogical Society of America). A bi-monthly publication in which each issue has a specific theme, addressed at a non-specialist level.

J. R. Craig and D. J. Vaughan. *Ore Microscopy and Ore Petrography* (2nd edition; John Wiley & Sons Inc., 1994). The most widely used book about reflected light optical microscopy and its use in the study of ore minerals.

J. R. Craig, D. J. Vaughan, and B. J. Skinner. *Earth Resources and the Environment* (4th edition; Prentice Hall, 2011). A comprehensive introduction to all Earth resources, particularly those studied in geology courses, along with discussion of the environmental impact of their exploitation and utilization.

M. Darby Dyar, M. Gunter, and D. Tasa. *Mineralogy and Optical Mineralogy* (Mineralogical Society of America, 2008). A comprehensive, modern core textbook covering all aspects of mineralogy and the optical properties of minerals studied in thin section.

W. A. Deer, R. A. Howie, and J. Zussman. *An Introduction to the Rock Forming Minerals* (3rd edition; Mineralogical Society of Great Britain and Ireland, 2013). The most comprehensive text covering the crystal structures, chemical compositions, phase relations, properties, and geological occurrence of the main minerals found in rocks.

C. Klein and B. Dutrow. *Manual of Mineral Science* (23rd edition; John Wiley & Sons, 2008). The most recent edition of the classic textbook of mineralogy, originally written by James D. Dana in

1848 and subsequently revised and updated numerous times with input from various new authors.

A. H. Knoll, D. E. Canfield, and K. Konhauser (eds). *Fundamentals of Geobiology* (Wiley-Blackwell, 2012). A comprehensive survey of the new field of geobiology with 22 review articles by experts in the field.

D. A. C. Manning. *Introduction to Industrial Minerals* (Chapman and Hall, 1995). An excellent introduction to mineral raw materials.

A. Putnis. *Introduction to Mineral Sciences* (Cambridge University Press, 1992). An excellent introduction to minerals studied in the context of processes rather than the systematics of mineral families.

J. Zussman (ed.). *Physical Methods in Determinative Mineralogy* (2nd edition; Academic Press, 1977). Despite its age, still an excellent introduction to the principal methods used to identify minerals.

Index

Index

Expand your collection of
VERY SHORT INTRODUCTIONS